# *The Formula Book 3*

# The Formula Book 3

Edward Nigh
and
Stark Research Associates

SHEED ANDREWS AND McMEEL, INC.
Subsidiary of Universal Press Syndicate
KANSAS CITY

## CAUTION TO READERS

In writing *The Formula Book 3,* I have excluded hazardous materials wherever possible, but in some cases they must be included to make a product effective. In making a formula, the reader should observe any note of caution added at the end of the recipe. In addition, cautions which are specific to an individual ingredient are indicated by an asterisk following that ingredient's name. The reader should consult Appendix B (Definitions of Chemicals) for the cautions pertinent to that ingredient.

We all know that materials such as waxes and oils will burn, so I have concentrated the warnings to materials that may be less familiar. But remember, *all* chemicals, including ordinary table salt, should be kept out of the reach of children, carefully labeled, and used only for the purpose they are intended.

The value and safety to you of the products in this book depend upon your careful use of the materials shown in the proportions given, as well as your observing any special cautions appearing in the book or with the materials. Neither I nor the publisher can be responsible for the efficacy of the products or your own safety if you do not follow these instructions and precautions.

 For information address Sheed Andrews and McMeel, Inc., Subsidiary of Universal Press Syndicate, 6700 Squibb Road, Mission, Kansas 66202.

**Library of Congress Cataloging in Publication Data**

Nigh, Edward.
The formula book 3.

1. Recipes. I. Stark Research Associates.
II. Title.
TX158.N48 640'.2 78-23188
ISBN 0-8362-2205-9 paper
0-8362-2202-4 cloth

# Contents

# Foreword

*The Formula Book 3* is the third in a series that will become a complete set over the coming years. These books are a text on the practical application of chemistry. They contain formulas of chemical compounds that constitute the many products used in our daily lives. We use such items so frequently and with such safety and effectiveness that we sometimes forget how dependent we have become on chemicals for our normal everyday functions.

It is the desire of the author that these formula books give the reader a greater appreciation of our dependence upon chemical knowledge. Perhaps there may be developed a better understanding of how our life processes are so completely dependent upon chemistry for survival. This understanding may create a greater desire for further study of chemistry, thus enabling one to better understand the implications of the science in today's technical world.

The formulas in *The Formula Book 3* have been selected because of consumer interest that has been expressed in many of the letters received by Stark Research Corporation requesting specific information for developing a particular product. Naturally, all such requests cannot be recognized, due to such factors as safety or complications in formulating the compounds or insufficient knowledge pertaining to the process of formulating a particular product.

It is hoped that the various formulas in the five sections of this book will give the reader sufficient products that he or she can formulate to make this third volume of practical and useful value.

Many of the suggestions received from users of the first two volumes have been included in this book. Great effort has been made to improve this one; however, there will still be formulas that are in demand which do not appear in this volume. Suggestions or criticisms from our readers are solicited in order to improve future editions.

# Introduction

Chemistry was one of the earliest sciences to be intensively studied, and in its continuing growth today is a part of our knowledge that is revolutionizing our way of life. It is an area of remarkable breadth and depth. It involves the study of change, usually with an action of one substance upon another that results in the formation of one or more completely different substances.

Earliest history records man's interest in such natural phenomenon as fire, lightning, and thunder, as well as common things such as air, water, moon, sun, and stars. For thousands of years he attempted to describe them in terms of simpler substances. Superstitions and black magic played a major role in his earlier attempts, with folklore and religious dicta frequently dominant.

The beginnings of chemistry go back to ancient times, in which practical technologies were developed for making metals and glass and producing lime and charcoal. Alchemy, developed by Arabic cultures, was concerned with the transmutation of such basic metals as iron, copper, and tin into gold. While gold was never produced, of course, its practitioners gave rise to the role of chemistry in the cure of disease. They discovered many chemical facts in their search for transmutation processes. In 1774, oxygen was discovered by Joseph Priestley, and followed by discovery of the true nature of combustion by Antoine Lavoisier. These discoveries began the modern science of chemistry, which has been developing to the present day.

One of the major reasons for the world population's doubling in the past 50 years, with anticipated redoubling in the next 30 years, has been the application of chemical knowledge in agriculture and medicine. The control of diseases through the use of drugs and antibiotics has been based in large part upon chemical research. Fertilizers and pesticides have played a major role in food production, necessary to feed the growing population.

Our food, clothing, and basic necessities are produced by processes that are chemically sophisticated. The pseudochemical jargon that is used in advertising everything from detergents to morning cereals makes us aware that possibly there are important chemical distinctions between materials that appear to be very similar.

Attention is now being given to ecology and the environment in which we live, bringing a wider concern for chemical questions by the general public. Many of our modern problems have chemical overtones which relate to the biosphere, or life zone, in which we exist. Such problems can only become magnified as the world continues to witness human population growth and pollution.

Chemistry has been called by some the "servant science" because it supplies descriptions and understanding of the many kinds of matter that are studied in detail in other sciences. In the study of natural science, chemical facts and theories are of great importance for scientists as well as nonscientists; a study of chemistry enriches their appreciation of the world of nature and even that "unnatural" world of new materials that is provided by the chemical industry. As public concern increases over issues that pertain to our general welfare, we must have the scientific truth required to make correct decisions for the benefit of all mankind. Perhaps this book may be an incentive for you to broaden your knowledge of chemistry with increased capacity to assist in making correct decisions. We will need all the help we can pull together for the future.

As you begin to use this book, some of the major concerns that will probably be encountered are what types of equipment do I need, where do I get the chemicals listed in the formulas, and how safe is the process of mixing the ingredients? These are certainly legitimate concerns, and it may be best to discuss them as individual topics.

## BASIC EQUIPMENT

All of the formulas listed in this book have been selected on the basis of simplicity in formulating them. With the exception of compounds which must be melted by heat, most of the processes involve a simple change in structure, such as a solid into a liquid by simple dissolution or by mixing two ingredients together, either as liquid or as solids in powder or granular form. Simple kitchen utensils are totally adequate in making over 95 percent of this book's formulas.

Several recipes require heating two separate ingredients in double boilers, so we suggest having two Pyrex double boilers. The glass pots have the advantage that one is able to see what is going on inside, and they are easy to clean. A good set of measuring spoons is also recommended, as all measurements in tablespoon or teaspoon should be made on the basis of one level container of the size recommended.

Since many of your finished products will end up in storage or left-over food containers, *always* label all containers — immediately after filling — with the name of the product and warning information, as appropriate. Keep all products, including any left-over raw products, out of the reach of children and animals. (A more detailed description of utensils and equipment useful for making the formulas is found in Appendix D.)

## WHERE TO OBTAIN THE RAW MATERIALS

Almost 90 percent of all raw materials in these formulas can be obtained from local sources, providing you have access to a town of substantial size. Drugstores, hardware stores, paint stores, fuel supply and lumber distributors should be able to supply most of your needs. Other supplies, such as chemicals, can be ordered from supply companies or others who distribute chemical supplies. This is discussed in more detail in Appendix C.

The selection of materials depends on the end use of the product. For example, selection of denatured alcohol for use in a paint or varnish thinner would be entirely different from selection of a product that would be in contact with the skin.

Two basic grades of chemicals are used: (1) USP and (2) "manufacturing" or "technical."

1. USP is an abbreviation for *United States Pharmacopoeia*, which is the official federal book of chemicals and drugs. This publication sets the standards of purity and other specifications that manufacturers must comply with. Generally speaking, USP grades are used in compounds that are taken internally or come in contact with delicate areas of the body that require pure materials. For example, when a material such as magnesium sulfate (epsom salts) is contained in a product for internal use, it must be USP grade.

2. In the manufacturing or technical grade category, the standards for purity are understandably less than in the USP grade, in that the end product does not directly affect human health. For example, if epsom salts were to be used in a foot bath, the purity requirements would not be the same as for internal use, and the manufacturing or technical grade would be acceptable.

There is a substantial difference in the cost of these two grades of materials; nevertheless, selection should *always* be made on the basis of the end use.

Odor and color, when used in a preparation, are largely a matter of personal choice, and usually have no effect on the function of the compound. For example, in the formula for face lotion, the perfume in the product has no effect on the properties it imparts to the skin. However, if perfume or color is desired, it must be of a type that is compatible with the compound it is to be used in.

Dyes and perfumes fall into three general categories: those that are (1) soluble in oil, (2) soluble in water or alcohol, and (3) suitable for use in an emulsion. Examples are as follows.

Type 1 is soluble in oil and used in liquids, semisolids, and solids that have an oil base, such as baby oil.

Type 2 is soluble in water and alcohol and would therefore be used in compounds such as face lotion, where the base is water and alcohol.

Type 3, an emulsion type, logically is used in emulsions.For each formula that requires a perfume or dye, the type is *specified* in the formula. But a word of warning: Perfumes and dyes are highly concentrated and should be used very sparingly.

Alcohols are widely used in many areas of chemistry, and especially in compounding formulas, such as those in *The Formula Book 3.* As a matter of fact, it would be hard to conceive being able to make many of the compounds without it. But based on the mail we have received from many teachers and students, there seems to be some confusion over the two primary types, that is, ethyl alcohol in its pure form and ethyl alcohol that has been denatured, or denatured alcohol. We hope this section will promote a better understanding for those who may not be completely clear on the subject.

Ethyl alcohol (ethanol, grain alcohol), $C^2H^5OH$ or $CH^3CH^2OH$, is a clear, colorless liquid having a melting point of -117°C. and a boiling point of 78.5°C. It is miscible in any proportion with water or ether, and is soluble in a sodium hydroxide (caustic soda) solution. Flammable, it burns in air with a bluish transparent flame, producing water and carbon dioxide as it burns. Its density is 0.789 at 20°C.

Absolute (anhydrous) ethyl alcohol is obtained by the removal of water. One process for accomplishing this is to react the water in the alcohol, with calcium oxide, and then distill the alcohol.

Ethyl alcohol is made by (a) the fermentation of grains and fruits, and also directly from dextrose, (b) by absorption of ethylene from coal or petroleum gas, and then water reaction, and (c) by the reduction of acetaldehyde in the presence of a catalyst.

Ethyl alcohol is used in tremendous quantities in beverages which are taxed by the federal government. There are many other uses as well, such as in pharmaceuticals, tinctures, and extracts for internal use, where it is not taxed as it is for beverages. For these uses, however, a special tax-free permit must be obtained from the Alcohol and Tobacco unit of the federal government. Permits of this type are available to educational institutions as well; however, for small-quantity use, buying a bottle of 95 or 100 proof vodka is much less complicated.

Denatured alcohol is ethyl alcohol (the same as used in beverages), except that it has been adulterated with other chemicals that make it unfit for beverage use, while retaining its other characteristics. Therefore, denatured alcohol is not taxed as pure ethyl alcohol is, making it very much less expensive.

There are two basic types of denatured alcohol, completely denatured (CDA) and specially denatured (SDA). The denaturants are specified by the Alcohol and Tobacco Tax unit, and depend on the end use of the alcohol. For example, a denaturant acceptable for use in alcohol to be used as an industrial solvent would be entirely unacceptable for use in a body lotion or mouthwash because of its toxicity and irritating properties. Therefore, the *type* of denatured alcohol must be chosen for the compound it's to be used in.

## SAFETY IN FORMULATION PROCEDURE

We have all seen movies of the "mad scientist," working in his laboratory with beakers and glassware tubing running in all directions and erupting steam and smoke. Invariably, there is a scene where something "goes wrong" and there is an explosion, with serious damage. These things can happen, but it's rare when they occur anywhere other than the movie theater.

Adequate warning is given in each formula when there is any danger connected with the raw product or the item when finished. In most cases such warnings pertain to direct toxicity of the material to unprotected skin. Precautions have been taken in selecting the formulas to ensure there is no chemical reaction that would emit serious toxic fumes or become explosive. *However*, most of the formulas could be toxic, and many fatal, if taken internally; therefore, correct labeling and adequate storage of your finished products are imperative. Commercial products are required to have warning labels — and for sane reasoning. To avoid any possible error and to enjoy full use of this book, follow this very simple rule of labeling and warnings.

Remember, all of these formulas have been tested and found to be effective if the ingredients called for are used and the mixing procedures are followed. Some of the products you make will be inferior to the commercial equivalent. This is usually the result of not being able to recommend a particular chemical, or the unavailability of it, because of federal or state regulations. For example, only small quantities of face creams should be made at one time due to the fact that no preservatives are added to prevent molds from developing. For those who wish to avoid such preservatives, this is no problem, but to others, the federal regulations on the availability and use of some preservatives can be annoying. It should be pointed out that such regulations are almost invariably made on the recommendations of competent chemists and biologists who recognize and consider both the advantages and disadvantages involved in their decisions.

# 1

# Around the House

## Around the House

Air Purifier, Wick Type
Ammonia Washing Powder
Antiseptic Detergent
Bacon Preservative
Brass and Copper Cleaner
Cheese Wax
Coffee Stain Remover
Cut-Flower Freshener
Dry Cleaning Fluid
Egg Preservative
Egg-Yolk Stain Remover
Enamel Cleaner
Glass Cleaner
Gold Polish
Ice-Cube Tray Release
Ink Eradicator
Ironing Aid
Ironing Scorch-Stain Remover
Jar Sealing Wax
Jewelry Polish
Laundry Gloss
Linoleum Paste Wax
Liquid Floor Wax
Metal Cleaner, All Purpose
Mildew Stain Remover
Oak Furniture Cleaner
Oil and Grease Spot Remover
Paraffin Oil Emulsion
Pine Oil Disinfectant
Soapless Rug Cleaner
Toilet Bowl Cleaner
Urine Stain Remover
Wall Cleaner
Window Cleaning Powder
Window Cleaning Spray
Woodwork Cleaner

Our domestic life is so dominated by chemistry that we fail to recognize or appreciate the wonders provided by this field of science. As we enjoy our everyday activities, we take for granted the miracles provided by the countless hours of research by scientific teams of chemists, biologists, physicists, and physicians.

Our first drink of water in the morning contains chemicals that protect our health. The morning pill, which may be taken for varied and sundry purposes, is based on the premise of sound chemical and biological knowledge. Most of us also begin the day by washing our face, and the soap we use, in itself, has always been a chemical marvel to me.

Perhaps most of us use soap without a thought that a complex chemical process occurs each time we use it. Have you ever asked yourself *how* or *why* it cleanses the skin? Since its chemistry is unique, it is hereby presented so that you may enjoy bathing more, with knowledge of the chemical actions in the process.

To make soap, we use oils or fats from plant or animal sources. We react these with water, acid, or alkali in a natural process called "saponification," which is the formation of glycerol and fatty acids. These separate as salts when alkali is used, aided by the addition of sodium chloride. These salts are the soap, and ordinary "hard" soaps are made from the sodium salts. When potassium chloride is used, the potassium salts produce the jellylike or "soft" soaps.

Both forms are soluble in water, and water, when agitated, reacts with the hydrogen and oxygen atoms of the water to form lather or suds. When natural "hard" water, which contains dissolved calcium or magnesium compounds, is used with sodium soaps in washing and cleaning, the soap separates as insoluble precipitates — the amount of precipitate depending upon the minerals in the water. This represents a loss of soap for cleaning purposes, less suds, and accumulation of the precipitate in the material being cleaned. Where hard water is a problem, synthetic detergents are recommended, since these compounds do not form greasy precipitates with the mineral in the unsoftened water.

Soaps function as cleaning agents by several methods. First, they lower the surface tension of water, which gives the water the opportunity to penetrate the article being washed; they act as an emulsifier that brings about the dispersal of the oil and grease; and they absorb the dirt particles in the suds, allowing them to fall free in the water.

So the next time you use soap, you can consider yourself something of a chemist. Turn on the bubble machine!

### *Air Purifier, Wick Type*

Commercial household deodorants come in a great number of good-smelling fragrances and do an excellent job of "masking" unwanted household odors that become troublesome at just the wrong time. Rather than wait for these to build up, the wick-type deodorant tends to absorb such odors as they develop. This prevention is frequently a better solution than trying to mask the smell of fried fish prior to the arrival of the bridge group.

Try making your own air purifier, and compare the results with the method or product you are presently using. Add 3 tablespoons (45 ml) of chlorophyll to 2 quarts (2 l) of denatured or isopropyl alcohol* and 2 quarts (2 l) of water. Then stir in 1 cup (237 ml) of formaldehyde* (37-40% solution). To use, pour in a bottle and insert a wick, leaving about an inch exposed.

After a few days' use, you may wish to cut off the exposed end and pull out another inch of wick. Kerosene lamps and their wicks are ideal and may be obtained at most hardware or department stores. *Remember,* do not use them for a lamp if you find this idea appealing.

### *Ammonia Washing Powder*

A lot of chemistry has gone into the development of many of our modern-day cleansers and washing powders. Their sale is big business, as evidenced by the amount of time and money devoted to their advertisement on TV — to me, always a good criterion of what is making money for an industry. They are

**Cautions which are specific to an individual ingredient are indicated by an asterisk following that ingredient's name. The reader should consult Appendix B (Definitions of Chemicals) for the cautions pertinent to that ingredient.*

effective, safe to use, and frequently pretty darn expensive, when you figure the cost on an area-use basis.

Back in our grandmothers' time, a lot of faith was placed in the magical cleaning power of ammonia or compounds containing this chemical. And by the tremendous sale of such products today, it would appear that their use for cleaning problems has persisted over the years. While they may come in more sophisticated containers than those of 50 years ago, they are essentially the same formulations and they are all but magical in their ability to remove dirt and grime from many fabrics and types of surfaces.

Those who would like to carry the pioneering spirit further than using one of the same compounds used by our forebears, may wish to formulate an ammonia washing powder. This can be done safely (and with little time) by mixing 2 cups (454 g) of powdered soap with 2 cups (454 g) of ammonium carbonate.* After thoroughly mixing, use at the rate of 2 tablespoons (30 g) to each quart (1 l) of water for general household cleaning chores.

### *Antiseptic Detergent*

Food-processing equipment, such as meat grinders, slicers, milk separators, and cooking equipment, need to be thoroughly washed with a good antiseptic detergent. The same is true of equipment, bedding, and eating utensils from hospitals and other facilities where the sick are treated. The bacteria and molds which contaminate such equipment may be responsible for serious health problems. In many states, health regulations require that specific types of detergents and hot water of designated temperature be used for washing hospital bedding and food-processing equipment. Such regulations, while sometimes bothersome and expensive, help reduce serious epidemics. Those who have traveled in foreign countries may have learned to appreciate our laws from the unpleasant experiences encountered where sanitation practices are ignored.

One of the major causes of gastric disturbances is a bacterium called *Salmonella*. When this character establishes itself

** See Appendix B for cautions pertaining to this ingredient.*

in the digestive system, it's pretty certain that the vacation fun is over. This bacterium becomes established as a result of inadequate food preparation, spoilage, and insufficient attention to the process of cleaning equipment.

A formula that can be used as a regular washing detergent whenever antiseptic action is required can be mixed in your own kitchen. It may be used in the dish water, clothes washer, and for cleaning all those pieces of food-processing equipment.

Simply mix together 2 cups (454 g) trisodium phosphate,* 1 cup (227 g) sodium bicarbonate, and 1 cup (227 g) sodium pyrophosphate in a bowl or ceramic crock. Use gloves to mix thoroughly and transfer to suitable glass containers which can be sealed against moisture. Label to include contents, date, and a warning *not* to be taken internally. Use at the rate of approximately ¼ to ½ cup per gallon of water, the rate depending upon the concentration desired.

### *Bacon Preservative*

Two things I like to smell around the campfire are the odor of strong coffee and bacon frying. Both are as natural to me as the smoke from the campfire, and not to have either would be like not having gone on a cookout or camping. There's a lot of stimulation in a good cup of black coffee, and bacon can make calories available that will be needed to get up the side of that mountain.

But have you noticed that, without good refrigeration, bacon may become moldy and develop a rancid taste? This can be especially true if your outing lasts for several days and the humidity is high during warm months. What had been looked forward to as a good breakfast becomes a disappointment if that happens. But it's simple to reduce or even eliminate the problem.

Next time you get ready to go on that camping or hunting trip, wrap a cloth which has been soaked in a solution of 5 tablespoons (75 ml) of white vinegar and 1 cup (237 ml) of water around the bacon. (Since most packages are hermetically sealed, you won't do this until the bacon is opened and exposed to the air.) Try this and enjoy your bacon longer — you'll be glad you did.

** See Appendix B for cautions pertaining to this ingredient.*

*Brass and Copper Cleaner*

I'm pleased to see the great number of utensils that are coming back to our shelves made from brass and copper. In my estimation, nothing adorns a kitchen more beautifully than cooking utensils made from these metals, shining and properly displayed. During colonial days, most of the pots and pans (besides a large iron cooking pot) were made from heavy hand-pounded brass or copper. And from the look of those on display in the museums, people spent a lot of time keeping them clean, since they are worn pretty thin in spots from a lot of rubbing or abrasive cleaners, or both. Wood ashes were used back then for cleaning purposes, which may have had something to do with it.

For those who are fortunate enough to own some articles made with copper or brass and are faced with the age-old problem of keeping it shiny, here is a formula that should spare you time and effort. To ¼ cup (57 g) of sodium bicarbonate, add ½ cup (113 g) sodium metasilicate and ¼ cup (57 g) of trisodium phosphate.* Use gloves to mix thoroughly, apply to the metal surface with a damp cloth, sponge, or brush. Rinse with warm, clean water and wipe dry with a clean towel.

*Cheese Wax*

The renewed interest in developing and preserving one's own food must be in part due to increasing costs and the desire of many to avoid additives. I receive at least a half-dozen letters a week from readers who request information for canning and preserving foodstuffs, and my stock answer is to refer them to their home economist in the Cooperative Extension Service, which has reams of data on freezing and canning and information on the nutritional aspects of the finished product. As the public information arm of your state, land-grant university, the CES is highly qualified to assist you in all your food preparation requirements.

Making cheese is becoming popular, and once it is finished, a protective coating of wax protects it from excess drying and from becoming moldy. Such coating is also recommended for

* *See Appendix B for cautions pertaining to this ingredient.*

those carrying cheese on camping trips. It reduces refrigeration requirements and keeps insects and soil from getting into it. The quality can be retained for several weeks in the field, if done properly.

All you need is 4 cups (908 g) of beeswax and 4 cups (908 g) of paraffin wax. Melt the two together in the top of a double boiler. When melted, tie a string around the cheese and dip the cheese into the wax several times, letting each coat dry before dipping again. If it is to be stored for prolonged periods, store in a cool, dry place.

### *Coffee Stain Remover*

One of the more pleasant things about having a formal coffee is the opportunity to get out your finest silver service and china and your best tablecloth and napkins. Too frequently we keep these beautiful items from everyday use for fear they may be broken or in some way damaged. And when we finally bring them out for display and use on that special occasion, we find ourselves apprehensive that something will happen to them. And, sure enough, someone fails to stop the spigot on the coffee server and a few drops of dark coffee manage to find their way to that antique and cherished linen tablecloth that was passed down from your mother's side of the family. While your immediate impulse is to yank it off the table and run to the kitchen for emersion into cold water at the sink, you smile and indicate to your guest that no harm is done. If you've had this experience before and find that removing the stain after it has dried is next to impossible, you can relax the next time this happens, with confidence that the following formula will remove coffee stains from most fabrics.

To 1 tablespoon (15 ml) of glycerin, add an additional tablespoon (15 ml) of isopropyl alcohol* and 3 tablespoons (45 ml) of water. Into this mixture, add 1 tablespoon (14 g) of ammonium chloride and stir until the mixture is dissolved. Saturate the spot with the solution and let stand for approximately 5 minutes, then rinse well with clean water. In most cases, you'll be ready for

* *See Appendix B for cautions pertaining to this ingredient.*

your next "coffee" and eager to invite the same guests. Try it; you'll be pleasantly surprised.

*Cut-Flower Freshener*

Flowers add that fresh and natural charm to any home that uses them for decoration. Whether they are purchased or brought from seedlings to flowering maturity in the home garden, they are cherished as long as their appearance has that look intended for them by Mother Nature. But when, after a few hours, they begin to lose their freshness, petals dropping and fragrance fading, is there some method to prevent losing our prized bouquet? There certainly is, and it is hardly farther away than your kitchen.

To ¾ cup of brown sugar (170 g) add 4 tablespoons of talc or kaolin (56 g), then mix with 2 tablespoons of yeast (28 g). To items 1, 2, and 3 add 1 teaspoon of pine or lime oil (5 ml), mix in 1 quart (1 l) of water, and use the mixture at the rate of 2 tablespoons to a vase (approximately pint size) of flowers, shaking before using. A small pinch of copper sulfate* may be added to retard mold formation if flowers are to be kept for a prolonged period.

This simple-to-make and inexpensive formula can add additional days to your enjoyment of cut flowers. Some types will respond better than others to the freshener, due to their ability to absorb the material.

*Dry Cleaning Fluid*

At the end of winter there is always an accumulation of wool clothing and blankets that have received their share of soil and perspiration. To have the entire lot dry cleaned is costly, and the convenience and economy of a do-it-yourself laundry may not be available to all. When you find yourself in this predicament and want to economize on the dry cleaning bill, you might try this technique.

* *See Appendix B for cautions pertaining to this ingredient.*

From your local oil distributor or chemical supply company, purchase a couple of gallons of Stoddards solvent.* Place the material in a shallow pan and agitate for a few minutes, paying special attention to particularly soiled areas. Wring out and air dry. Since the solvent is moderately flammable, *caution* should be practiced in using it. Keep away from open flames and store in a leakproof metal container, with the contents clearly labeled. Avoid prolonged inhalation (as with any petroleum product).

### *Egg Preservative*

Ever wonder why eggs spoil when they come wrapped in such an impervious shell? Well, the answer to that question is they are not completely impervious and, with time, air penetrates the shell and spoilage occurs. When it is necessary to store eggs for extended periods of time and cold-storage facilities are unavailable, you can use the old standby of adding egg preservative to prevent the air from penetrating.

Use the liquid form of sodium silicate* (waterglass) at the rate of 1 quart (1 l) to 2 gallons (8 l) of water. Boil the water first, cool, and then add the sodium silicate. Put the eggs in a crock or other large container and cover them with the solution. Keep them covered in a cool, dark place.

### *Egg-Yolk Stain Remover*

My introduction to the fine arts began in the third grade of elementary school when the teacher presented pictures of the master artists and their paintings. While I was impressed at the time, I must admit that even today my recollection of this encounter consists of the fact that egg yolks were one of the principal ingredients in the raw materials used in the paintings. I always wondered how they kept their pictures from smelling, or if they maintained a flock of chickens near their studios for a

* *See Appendix B for cautions pertaining to this ingredient.*

fresh egg supply. It's not too surprising that they would have gone to egg yolk since there is oil in it which today finds widespread use in our huge cosmetic industry. While many of us may not intentionally be using egg yolk for painting, we may be presented with the problem of removing it from articles of clothing, tablecloths, or woodwork.

A formula that will remove egg-yolk stain can be made by combining 1 tablespoon (15 ml) glycerin, ¼ cup (59 ml) isopropyl alcohol* and ¼ cup (57 g) powdered soap. Rub the mixture into the stained area and allow it to stand for 5 minutes. Rinse with clear water, and the fabric should dry without a stain.

### *Enamel Cleaner*

Enameled surfaces require care when it comes to cleaning persistent spots and stains. It must be remembered that the finishes are obtained through the application of paints, and these can be scratched or their lovely luster marred by the use of abrasives.

With some spots, such as badly burned foods, I frequently find it's most convenient to use one of the commercially prepared materials for cleaning ovens. But for ordinary, run-of-the-mill cleaning jobs, I use a formula that seems to work on all but the worst stains. It's simple, efficient, and there is no damage to the enamel.

To make this cleaner in your kitchen, you'll need the following chemicals, mixed as described, and you will be in business to clean those enameled pots, pans, or other utensils.

Add 3 tablespoons (42 g) of sodium carbonate and 1 tablespoon (14 g) of sodium metaphosphate to 2 tablespoons (28 g) of soap powder. Thoroughly mix with 1 cup (227 g) of fine pumice powder, and when all four ingredients are blended apply with a damp rag or sponge. Scour the surface and rinse with clear water. The results should please you, and you'll be pleased with yourself as well for the economy involved.

* *See Appendix B for cautions pertaining to this ingredient.*

## *Glass Cleaner*

Glass articles that become tarnished or lose their normal, clean appearance may frequently be suffering from particles of matter adhering to the surfaces. These vary, of course, depending upon the article and the purpose for which it has been used. A surface of cut glass, for example, may have accumulated a layer of dust and everyday debris from setting on a shelf or in a display case.

Whatever the cause, a cleaner that will bring a sparkle back to those cherished glass pieces can be made in a few minutes, and will save time in the cleaning process. Mix 1¼ cups (284 g) of calcium carbonate, 2 tablespoons (28 g) ground quassia, and 2 tablespoons (28 g) ammonium carbonate* together. After they are thoroughly mixed, apply with a damp cloth or sponge. "Cut" surfaces may need application with a small brush (an old toothbrush serves the purpose quite well). Rinse with clear water and wipe dry with paper towels.

## *Gold Polish*

While gold doesn't rust, it does oxidize and become tarnished. When this occurs, it is necessary to use a polish that is capable of bringing back the luster and soft rich color that only this highly prized metal has to offer. It doesn't require a lot of time and effort if you have a product that is effective. One which I find fills all requirements can be made by combining the following chemicals.

Mix together with fork or wooden spoon 1 cup (227 g) of Fuller's earth, 1 cup (227 g) of calcium carbonate, 2 tablespoons (28 g) of ammonium sulfate, and 1 tablespoon (14 g) of aluminum powder. Put in a suitable storage container, label well, and use by picking up a small amount with a soft damp cloth and rub it on the gold to clean and polish. Rinse with clear water and dry with a soft cloth.

** See Appendix B for cautions pertaining to this ingredient.*

*Ice-Cube Tray Release*

There are a lot of us who still use ice trays and suffer trying to get the cubes out of their containers after they have frozen. It can get downright exasperating, fighting a tray of ice cubes all over the kitchen while your guests sit patiently waiting for the drinks you promised.

A very simple solution is to mix corn oil and peanut oil at the rate of 1 tablespoon each (15 ml) and apply a very light coat to the trays and dividers. Rub off all excess with a cloth, and do not wash in hot water. You'll find one application lasts for several freezings. It sure helps you keep your cool in more ways than one.

*Ink Eradicator*

There were certain advantages to growing up during the Depression. Since money for recreational purposes was limited, one had to tax one's mental facilities to come up with bright ideas to keep oneself amused. The old games of Kick the Can, Annie Over, and Hide and Seek were some of the standbys which, apparently, our young generation of today does not understand. And there were always various clubs that were structured for different events, often depending upon the theme of the weekly movie serial at the local theater. It seemed that at least once a year a secret club would arise and invariably call for an ink eradicator, so that no one could read what had been written in a mysterious code.

If we ever got any of the messages eradicated, it was through simple destruction and not through the various chemical processes we dreamed up, which were always ineffective. But we survived and thought we were having lots of fun. I'm not sure how many secret clubs there are nowadays, and it would seem that messages are often sent on tapes; however, where an ink eradicator may be needed, here's a good formula that works well.

Mix 3 tablespoons (42 g) of alum with 3 tablespoons (42 g) of citric acid, then dissolve the mixture in 5 tablespoons (75 ml) of water. Apply a few drops and allow to stand for 5 minutes. Then use a little more, if necessary.

*Ironing Aid*

As you're ironing, you sometimes find the iron will not glide smoothly across the fabric. This is caused by particles of foreign matter that were not removed from the garment in washing, or from starch sticking to the iron, or both. You can prevent this problem if you take a few minutes' time to mix and use this formula. Try it and you'll find that ironing, while never fun, can be less troublesome.

Stir 1 teaspoon (5 ml) of silicone oil emulsion into 1 quart (1 l) of water. Spray this onto a cool iron or rub a cloth which has been soaked in the solution over the bottom of the iron. Heat until the water steams off, and the iron is ready for use.

*Ironing Scorch-Stain Remover*

Only a few years ago, the old flat iron was the standby for household ironing chores. They were placed on the woodstove, and when the temperature was thought to be about what was needed, they were removed and tested for the desired properties. Unlike today, exact temperature regulation was unheard of, which frequently resulted in scorching the garments. Even with our modern appliances, we still find ourselves overestimating the heat required to provide a smooth ironing surface, which results in a yellow to dark brown stain that may render the garment worthless.

Many times, depending upon the composition of the material and severity of scorching, these spots can be removed and the garment can be saved from the category of dustrag and made a usable piece of clothing once again.

Next time you're confronted with this unhappy experience, try this formula to remove the scorched stain — and with the money saved by not having to buy a new garment to replace the stained one, why not treat yourself to something new in wearing apparel the next time you go shopping? You owe it to yourself, really, for the economy involved.

Mix ½ cup (118 ml) of hydrogen peroxide (standard solution —3%) and ½ cup (118 ml) of water. Soak the scorched spot in the mixture for 10 to 15 minutes, then rinse in clean water for approximately 1 hour. The stain can usually be removed from fabrics that are made of natural fibers. The treatment is less effective with synthetic fibers, but it is always worth trying when a favorite blouse or shirt is stained.

*Jar Sealing Wax*

Since there has been renewed interest in canning in the past year or two, a great number of books, brochures, and general information have become available to the public — some of it good and some of it bad. And the interesting thing is that there isn't that much involved in the operation. If you follow the simple directions, it's fun — or not really difficult — to "put up" most vegetables, and certainly a way of economizing if you have bought the produce at a reasonable price. One of the best sources of information is in the office of the home economist of the Cooperative Extension Service, which makes published information available and frequently puts on demonstrations for the public. The CES can usually be located under the name of the county in which you live, or the name of your land-grant university which is frequently better known as the College of Agriculture.

The most important step in canning is to sterilize, for assurance that there will be no spoilage. This, of course, can be quite dangerous if not done properly, as botulism poisoning can result with some food materials.

When you have completed the process and are ready to seal the containers, you may find that paraffin wax is recommended as a sealant, especially for preserves and jellies. To obtain a tighter seal and better protection for your labors, it's a good idea to consider this formula the next time you make your own preserves or jellies.

Melt 1 pound (454 g) paraffin wax with 4 tablespoons (37 g) of stearic acid in the top of a double boiler. When melted, pour the mixture over the top of the preserves. The stearic acid will increase the density of the wax, which produces a tighter seal. Simple, but effective.

*Jewelry Polish*

Most jewelry is intentionally made ornate, both for beauty and as evidence of the artistry of the one responsible for making it. As a consequence, no thought is given to the problem of cleaning a fine piece of gold, silver, or one of the many other less precious metals which are used in production of today's wide assortment of jewelry. As time and the elements have their influence on your favorite pieces, you may be wondering if a stop at the jeweler's isn't the order of the day, if not to get them cleaned, at least to purchase some of the "magic formula" they have which puts that new gleam into the old metals.

But why buy it when a few simple chemicals may be combined with the end result being as effective as the much more expensive "store bought" product? You can formulate your own polish by following the simple recipe which follows, enjoy the fun of making it, save money while doing it, and end up with a sparkling product for your efforts.

To 5 tablespoons (70 g) of ferric oxide mix 2 tablespoons (28 g) of calcium carbonate, then add sufficient tap water, while stirring slowly, to make a paste. Apply with a cloth or toothbrush (or both if the need calls for it) and wash with warm water. Dry.

*Laundry Gloss*

Ironing is a household responsibility that can best be accomplished by a commercial laundry, as far as I'm concerned. But it's expensive, and the inconvenience of carrying everything back and forth is a hassle. Besides, many of us live in areas that are inconveniently located to such facilities. Therefore, we find that most of our ironing is still done at home, but many commercial products have been made available in the past decade which help reduce ironing chores; so there are brighter sides to the picture.

One product that can make your clothes look brighter and ironing more easy can be made at home. It improves the starch and provides a smoother finish for ironing. Once you've made it, you'll have a supply that will last you through many weekly ironings.

Soak 1 tablespoon (14 g) of gum acacia in 1 pint (474 ml) of water overnight. Then add 1 ounce (28 g) of borax and heat to a boil, add 1 tablespoon (15 ml) of glycerin, and cool and strain. Use by adding 1 tablespoon to 1 pint of starch.

### *Linoleum Paste Wax*

As a kid, one of the more unpleasant tasks I was assigned was washing the kitchen floor and waxing it following the weekly clothes-washing chores. The responsibility was undoubtedly good for me, but I resented the time it took from my other interests.

I don't know what type of wax was used in those days, but it was difficult to apply and required a lot of rubbing to bring the desired gloss. The only fun in the whole operation was putting my younger sister on a wool blanket and sliding her over the surface to polish the linoleum. This worked well, until I got carried away and practically slid her through the kitchen door while it was closed. From then on, a cloth-rubbing technique was imposed on my operation.

Today, through the miracle of chemistry, industry has produced outstanding floor polish and waxes. Those for linoleum-covered floors are numerous and efficient and of the type I probably dreamed of having during my floor-waxing days. Those who may not have the commercial products available may wish to consider formulating a linoleum paste wax that puts a shine and hard finish on the floor that will last a surprisingly long time.

Melt 1½ cups (340 g) of carnuba wax with 1½ cups (340 g) of ceresin wax in a double boiler. Remove from the heat and allow to cool for several minutes. Then slowly stir in 4 cups (948 ml) of mineral spirits.* As the mixture begins to solidify around the edges, pour into containers. The wax can be applied with a dry cloth or sponge. When dry, polish with a soft cloth or buffer.

### *Liquid Floor Wax*

I'm not sure how to get a genie from a bottle of floor wax, but I

* *See Appendix B for cautions pertaining to this ingredient.*

do know how to formulate a product that seems to work about as well as some of those commercially advertised. Once you've gone to the trouble of getting the old wax removed and the floor clean, you're ready for a product that not only will look good, but also serves as a protector of those beautiful wooden floors. You can make a quantity in a short time with minimum effort. The ingredients are all available at your grocery or paint stores, and hobby shops also carry them.

Melt ¼ pound (57 g) of beeswax and 1 pound (454 g) of paraffin in the top of a double boiler. Remove from the heat and, as it cools, stir in ½ cup (118 ml) of raw linseed oil and 3 cups (711 ml) of turpentine.* Stir rapidly until all of the ingredients are well mixed. To use, apply a thin coat to the floors, allow to dry, and polish with a clean cloth or electric buffer.

### *Metal Cleaner, All Purpose*

Cooking utensils, metal serving trays, wall hangings, ornaments, auto parts, household implements, window frames, screen doors, flatware, and drinking utensils are but a few of the items used in the average household which are made with various types of metals. Wear and tear, and frequently the elements, may remove the original look and reduce appearance to the point where the items are unacceptable for further use. If you encounter this situation, mix the following formula and use it a few times before you make the decision to dispose of the items. Frequently you can bring back the original luster, and the metal will sparkle like new when this cleaner is used.

In 1½ cups (341 g) of trisodium phosphate* mix 6 cups (1.4 kg) of soda ash and 2½ cups (567 g) bicarbonate of soda. (Gloves are recommended when mixing.) Mix thoroughly, using a fork or wooden spoon, and apply with a damp cloth or sponge. Rinse with clear water.

### *Mildew Stain Remover*

Fungi are both friend and foe of man. Some of our most

** See Appendix B for cautions pertaining to this ingredient.*

serious human diseases are caused by different species of these primitive plants, while others are capable of causing diseases in our cultivated crops. The infamous Irish famine was the result of a disease in the potato which completely eliminated the yield in many parts of Ireland. Hardly any plant or their products are spared from the attack of one or more species of fungi. The deterioration of clothing, house paints, wood, leather, and even the rubber in the tires on our automobiles may be attributed to microbial action.

One of our first and most useful antibiotics, penicillin, is a product of a fungus. Other species produce many useful products of medicinal importance. The multimillion-dollar mushroom industry in this country is testimony of our taste for this form of fungus. Soy sauce, used for seasoning, is a by-product of fungi. Certainly the list of products used to our advantage is large.

One fungus product we do *not* find to our advantage is mildew, and we frequently encounter a problem with it in our fabrics when excess humidity is present. A simple way to reduce the problem is to eliminate the humidity. However, where this method isn't possible, the resulting stains must be removed if the particular article is to have additional use.

A simple formula that can be made both easily and inexpensively at home will eliminate or help reduce mildew stains. Though commercial materials are available, you'll find satisfaction and savings in preparing your own formula in a few minutes' time.

To formulate a mildew stain remover, mix 4 tablespoons (60 ml) of hydrogen peroxide (standard solution—3%), 1 tablespoon (15 ml) of ammonium chloride, and 2½ tablespoons (37 ml) of denatured alcohol* or isopropyl alcohol* into 1 cup (237 ml) of water. Use by soaking stained areas in the mixture for 5 minutes, rinse well with clear water. Repeat if necessary.

### *Oak Furniture Cleaner*

Solid oak antique furniture is almost priceless today. If you have any doubts about that statement, drop in on your nearest

** See Appendix B for cautions pertaining to this ingredient.*

antique dealer and price some of the old round dining tables that weigh a ton, or an antique rolltop desk. The demand has been so great for oak furniture in a beautiful natural finish that manufacturers are answering the call with oak furniture just as exquisite as that made 100 or more years ago.

Anyone fortunate in having furniture made of one of our best native hardwoods finds that an occasional cleaning is in order. For ordinary cleaning purposes, one of the commercial products that comes in a pressure atomizer may be entirely satisfactory. But when dirt and grime have accumulated over the years, a real scrubbing may be in order. If so, try using ½ cup (113 g) of soda ash in 1 quart (1 l) of warm water. Use a soft brush and rinse with a cloth moistened with clean water. Dry and apply your favorite oil or polish. It's also a good way to prepare a surface for refinishing.

### *Oil and Grease Spot Remover*

I'm one of those do-it-yourselfers who has to get thoroughly immersed in what I'm doing to be anywhere near successful. For example, if I'm painting, it's rather difficult to determine if the article has more paint on it than I have on me. And the same may be said if I attempt to change the oil and filter on the family car. It seems impossible to accomplish these simple tasks without looking as though I'd changed a complete engine. And sometimes the necessity to do these chores strikes when one is not dressed for such activities, resulting in oil and grease spots on that good jacket or sweater you received as a Christmas present. This could require serious explanations; however, you can formulate an effective solution to remove such spots, and with a little time and effort, no one will be the wiser for the mishap.

Use a spoon to mix 1 cup (454 g) of sodium aluminate in approximately 2 quarts (2 l) of water. Before using, warm the solution and immerse the soiled area. For small spots, you can place a blotter under the area and apply the solution with a sponge. After drying, wash the garment normally. While most fabrics are colorfast, you might try a few drops on an isolated area first to make certain there will be no discoloration from the solution.

*Paraffin Oil Emulsion*

Tile floors were once popular and in demand, especially in the Southwest, where the decor was part of the Spanish-Mexican influence. They are still used to a limited extent but their cost and competition with excellent linoleum floor coverings has reduced the demand in all but the most expensive homes.

Those who are the happy owners of homes with tile floors sometimes doubt their good fortune when it comes time to clean and polish them. A bright and shiny surface isn't desired, but something to penetrate the Mexican or quarry tile to give it a new and natural appearance is required for a beautiful floor.

I happen to be a happy tile-floor owner and I have found a simple and effective method of taking care of it. I use the formula but once a year on all but the most traveled areas. Compliments indicate that the method is quite satisfactory, and it is inexpensive as well.

In a double boiler, heat 2½ cups (592 ml) plus 5 tablespoons (75 ml) of oleic acid, 1 tablespoon (15 ml) of triethanolamine,* and 3 cups (711 ml) of water. After heating (do not boil), beat with a whip or electric mixer until a milky emulsion forms. To use, apply with a sponge or cloth to the floors. Rub off with a clean, dry towel after letting set overnight.

*Pine Oil Disinfectant*

Until I watched the TV commercials, I hadn't realized there are so many places where germs can multiply. It seems they lie in wait for some unsuspecting soul to come by, then wreak havoc with their health. We know that bathrooms are a prime area of invasion because of moisture, which is required for multiplication of bacteria and fungi. Around the drains of sinks, tubs, and showers, a good shot of disinfectant will help keep them clean and odorless. Floors should be wiped a couple of times a week

** See Appendix B for cautions pertaining to this ingredient.*

since they are usually damp from bathing. If the bathroom is carpeted, it is a good idea to apply a little of the same material as a fine spray.

Besides the bathroom, the other area of the house where sanitation is most important is in the kitchen. Germs can collect around and under the sink. Being close to food preparation areas, it should be kept as clean as possible for sanitation purposes. Floors should be cleaned often; they too can use a disinfectant that will keep germs from invading our food. Nor does it hurt to use the same treatment around the area that has been assigned the family dog or cat for sleeping quarters.

A product that I've found effective — and that has a natural, refreshing odor — uses pine oil. It's just as effective as the kinds you see advertised, and making it yourself gives a lot of self-satisfaction when you see the advertisements on TV. It can be easily made in a few minutes' time by following this simple recipe.

To 2 cups (474 ml) of raw pine oil, mix 1 cup (237 ml) of sulfonated castor oil and 6 tablespoons (89 ml) of oleic acid. When thoroughly mixed, add 1 teaspoon (4 g) of caustic potash.* To use, add about ½ cup to a gallon (4 l) of warm water — for mopping and spraying into corners and drains. *Caution* should be practiced with the caustic potash, as it can cause skin burns; so handle with care. If it gets on your skin, rinse with water immediately.

### *Soapless Rug Cleaner*

Next time you find spots on the rug or carpet, you can do one of two things: call a commercial rug-cleaning firm or attempt to clean it yourself. Unless the job is too large (either in size or because of the material which caused the spot), it's less expensive to follow the second alternative. If you do, consider one of the very excellent commercial products on the market, expressly for the purpose of spot removal. But if you're caught and time is essential in getting the spots removed, you can make your own soapless rug cleaner that will work on most fabrics and on a wide range of substances responsible for spotting rugs and carpets. It's

* *See Appendix B for cautions pertaining to this ingredient.*

an inexpensive formula and takes but a few minutes to make.

Stir 1 cup (237 ml) of white vinegar into 1 quart (1 l) of denatured alcohol* or isopropyl alcohol*. (Vodka may be used for denatured alcohol in real emergencies.) Apply this to the rug with a sponge-head mop and rinse with clean water. Repeat if needed.

### *Toilet Bowl Cleaner*

Perhaps one of our unsung heroes of history is an Englishman by the name of John Crapper, who invented our modern toilet facility and thereby advanced sanitation engineering into a new era. We take for granted our modern conveniences, but have only to visit abroad or read of the epidemics that still plague many developing countries because of inadequate sanitation facilities to be reminded of our good fortune. Having been in some of these countries and witnessed the miserable and unhealthy conditions under which the people live, I think it's appropriate we pay due respect to John and others, such as Louis Pasteur, who have been responsible for doing so much to reduce human sickness.

To keep the toilet bowl sparkling clean and sanitary requires attention and a little effort about once a week. The type of local water usually dictates how frequently our efforts are required. Water with high concentrations of calcium or iron tends to stain or leave rings on porcelain and the need for cleaning may be more frequent. While there are many products on the market that will clean your toilet bowl, you can easily prepare a mixture from the following formula that will be just as effective and much less expensive.

In 1½ cups (340 g) of caustic soda,* mix ¾ cup (170 g) of powdered alum and ¼ cup (57 g) of sodium chloride. Mix well together with a wooden spoon and store in airtight containers. To use, wet the bowl sides by flushing or splashing water with a brush, then sprinkle the mixture in the bowl. Leave for at least 10 minutes then scrub with the brush and flush once more. Your final product — a clean bowl that you can be confident is a compliment to our sanitary engineering advances.

** See Appendix B for cautions pertaining to this ingredient.*

A word of *caution:* caustic soda can cause burns and should be handled carefully. Use gloves and flush with clear, cool water if a stinging or burning sensation appears on any part of the skin due to contact with the soda.

*Urine Stain Remover*

The number of "companion" animals is increasing in the United States and urbanization finds many more pets living in our homes. They are close to us, and soon become members of the household. Because they spend more time within the home, the chances are greater that even the best-trained, house-broken pet will sooner or later have an unavoidable accident. And what about that puppy? Mopping up becomes a chore! In spite of all the children's promises to take care of such matters, we are invariably confronted with stains on our best rugs . . . in the most conspicuous part of the living room carpet.

While there are times when you have to count to ten to avoid major conflicts over such issues, keep the following in mind and it will assist in reducing the burden, as well as the cost of removing urine spots.

A simple method that works on most fabrics, including the majority of the synthetic fibers in today's carpets, is to use a 10% citric acid solution. Saturate the stained area with the solution for approximately 5 minutes (depending on how long the spot has been there). Wipe off and rinse, or wipe with a sponge soaked with clear water.

For *persistent* spotting problems, those who have little experience in house-training a pet may wish to follow the advice in many good books on the subject, or your veterinarian can make suggestions. But until that time when you can relax and have the fullest confidence in that little bundle of joy in the fur coat, I'd suggest you keep the bottle of citric acid on hand — labeled and out of the reach of children.

*Wall Cleaner*

The prospect of cleaning walls in the ordinary home can be

overwhelming when one contemplates the square footage. Kitchen walls are frequently a chore to clean, since they are more subject to accumulation of grease, while bathroom walls become spotted due to moisture from showers and baths. The prospect of having to clean such areas can bring on headaches and the thought of tired muscles. Perhaps it will ease the burden if you have a good wall cleaner available, and especially one that isn't too expensive. You should find soda ash on the soap shelves of your favorite market, and in garden supply stores you'll find the second ingredient, ammonium sulfate.

In 1 cup (240 g) of soda ash, mix 1 tablespoon (15 g) of ammonium sulfate, and use at the rate of 2 tablespoons (30 g) per quart (1 l) of water. Apply to walls by sponge or cloth and wipe with cloth dampened with clean water. Hopefully, all your headaches will disappear with minimum effort.

### *Window Cleaning Powder*

Air pollution continues to cause problems for our health and welfare, and one of the less important problems (a by-product of the condition) is dirty windows. Particulate airborne matter adheres to the glass, especially when it is wet from rain or when water evaporates due to the difference between the inside and outside temperature. As a result, a film develops on the glass and, depending upon your geographic location, the film becomes a chore to clean. In the major cities, where pollutants from motor vehicles are high, a window cleaner may be needed that is stronger than the usual commerical types, which are principally alcohol based. Those who find ammonium products undesirable or ineffective for particular cleaning problems may wish to try the following formula, which I have found highly satisfactory for some of the most obstinate jobs.

In 1¾ cups (398 g) of borax mix 3 tablespoons (42 g) of sodium metaphosphate. Store in a suitable container and label. To use, simply apply with a damp cloth or sponge. Rinse with clear water or a clean, wet rag.

You'll be pleased at the sparkle that results, and surprised at how *dirty* your windows had become since they were last thoroughly cleaned.

### *Window Cleaning Spray*

Windows become dirty during the winter months due to the snow and rain; pollutants in the atmosphere cause spots and stains. Inside, condensation keeps the glass wet and causes streaks and spotting. Regardless of *how* they become dirty, it all adds up to work and expense if you have a lot of windows and buy commercial window-wash products.

I'm not sure that I have any good suggestions for keeping water off the outside of your windows. Storm shutters help, but they too become soiled. Condensation inside can be reduced by minimizing moisture in the home and keeping the temperature at a reasonable level. The next best thing to do is start washing when the weather permits; but instead of buying window cleaners, try making your own. The money you save will make the chore a little easier. It's simple to make and requires but a few minutes to mix. The ingredients are readily accessible.

Use 1 cup (237 ml) of isopropyl alcohol* and 2 cups (474 ml) of water. Then stir 5 drops (1.5 ml) of acetic acid or 1 tablespoon (30 ml) of vinegar into the mixture. (Though it won't add to its effectiveness, you can add a drop or two of your favorite food coloring. Blue is the usual color in commercial cleaners; so try it for its psychological effect before trying another color.) Place the mixture in a container with a spray attachment and you're in business.

### *Woodwork Cleaner*

Because lumber prices have soared, architects are using less wood in homes today than ever before. I personally hate to see it, as the warmth of natural woods is always beautiful and appealing. Those of you who own older homes, where wood was used for interior decoration and utility, are indeed lucky to have some of these masterpieces. Someday, perhaps, we may be able to return to such practices, but I doubt it. In spite of increased timber production through forest genetics and management practices, the demand for housing will require most of our lumber resources for structural use and not for interior finishing.

** See Appendix B for cautions pertaining to this ingredient.*

Like any other, wooden surfaces become soiled and require cleaning. Several efficient commercial types are available in liquid and aerosol form. Where a lot of washing is required, such products may be costly. Besides, you may not always have them on hand; so if you don't, or wish to formulate a product that can be equally effective, try this simple recipe.

Mix 1½ cups (355 ml) of oleic acid with ½ cup (118 ml) of triethanolamine.* Add 2 tablespoons (30 ml) of this mixture to 1 gallon (4 l) of warm water. Apply to surfaces with a cloth or sponge. Where grease has become a real problem, you may have to double the concentration. Wipe with a clean, wet cloth or sponge to remove soiled residues released by the cleaner. This should bring back the natural finish to wood with little effort — and less expense.

* *See Appendix B for cautions pertaining to this ingredient.*

## NOTES

Name of Formula: ______________________________

Date Made: ____________________

Ingredients and amounts: ________________________

______________________________________________

______________________________________________

______________________________________________

Label: Ingredients and caution warnings

Observations: ____________________________________

______________________________________________

______________________________________________

______________________________________________

______________________________________________

----------

Name of Formula: ______________________________

Date Made: ____________________

Ingredients and amounts: ________________________

______________________________________________

______________________________________________

______________________________________________

Label: Ingredients and caution warnings

Observations: ____________________________________

______________________________________________

______________________________________________

______________________________________________

______________________________________________

## NOTES

Name of Formula: ______________________________
Date Made: ____________________
Ingredients and amounts: ______________________
______________________________________________
______________________________________________
______________________________________________

Label: Ingredients and caution warnings
Observations: _________________________________
______________________________________________
______________________________________________
______________________________________________
______________________________________________

- - - - - - - - - -

Name of Formula: ______________________________
Date Made: ____________________
Ingredients and amounts: ______________________
______________________________________________
______________________________________________
______________________________________________

Label: Ingredients and caution warnings
Observations: _________________________________
______________________________________________
______________________________________________
______________________________________________
______________________________________________

# 2

# Personal Care

## Personal Care

After-Shave Lotion
After-Shave Talc
Antichap Handcream
Antichap Stick for Lips
Athletes Foot Ointment
Brilliantine
Brushless Shaving Cream
Contact Lens Cleaning Fluid
Cuticle Cream
Dandruff Treatment
Denture Adhesive
Denture Cleaner
Face-Powder Base
Fingernail Bleach
Fingernail Polish Powder
Footbath Powder
Greaseless Hair Dressing
Hair Cleaner
Hair Shampoo
Hand Lotion
Ink Remover for Hands
Lime and Glycerin Cream
Makeup Remover
Mouthwash Powder
Nourishing Cream
Softening Bath Water
Sore Muscle Rub
Tooth Powder
Vapor Inhalant

The history of cosmetics probably dates back to early men and women who used natural products such as soils of different colors, various stains from plant products, and carbon from burned matter to cover their bodies. This served the purpose of being able to hide from their enemies, or gave them the advantage of being able to slip up on them. The same pertained to their intended meal. The practice of camouflaging oneself apparently led to the use of the same materials for body adornment to call attention to oneself or a particular portion of the body. In the most primitive tribes known today, such practices are clearly evident. In fact, cosmetics play an important role in their social structure and tribal custom. Few religious ceremonies of any type are without some use of cosmetics, be it nothing more than the mystery involved with the exciting and perhaps strange smell of a plant or animal product which we designate as incense or perfume.

The use of cosmetics may be a sign or indicator of the development of civilization since it is possible to trace the periods of cultural growth by the widespread practice of personal adornment. Over 4,000 years ago the ancients were mining cinnebar for rouge. Malachite found favor around the eyes of the Egyptians. Henna was used not only for hair but also to redden the palms of the hands and the soles of the feet. The Greeks and Romans were noted for their excesses in grooming. Oils and fragrances were important in the rituals of the major religions. The word *perfume* is derived from the Latin: from *per* meaning "through" and *fumum* meaning "smoke," since the earliest known use of perfume was in the creation of incense.

The glories of grooming were never more pronounced or, perhaps as excessive, as in the courts of Europe when the Crusaders, returning with spoils from the conquered, brought cosmetics — until then unknown or unavailable to the European privileged classes. Before the Puritan era, efforts at self-beautification were nothing short of spectacular. Later, with the advent of the Victorian era, it has been said "a somewhat jaundiced eye was cast on such frivolities" — certainly an understatement when you consider the following law passed by the British Parliament during the reign of King George III in 1774:

"All women, of whatever age, rank, profession or degree, virgins, maids, or widows, that shall from and after this act impose upon, seduce and betray into matrimony any of His Majesty's subjects by the use of scents, paints, cosmetics, washes, artificial teeth, false hair, Spanish wool [a kind of rouge], iron stays, hoops, high-heeled shoes or bolstered hips, shall incur the penalty of the law now in force against witchcraft and like misdemeanors, and that the marriage, upon conviction shall stand null and void."

As someone has said, "You've come a long way, Baby!"

### *After-Shave Lotion*

My interest in after-shave lotion rose in direct proportion to my interest in girls. But back in those days, it was simple to make a selection since the total consisted of no more than four or five "smells." None was too strong and the scent didn't last long. Perhaps that's why I didn't get married until late in life.

Today, the sale of men's cosmetics is increasing as never before. Millions of dollars are spent for products designated as after-shave lotions. Most serve as an astringent and mild disinfectant, but they are really employed for the lasting fragrance that follows their use. Some are "masculine" while others are tailored for the feminine market.

If you'd like to break away from the usual, why not mix your next batch of after-shave lotion to your own specifications. Develop a fragrance that suits your personality. Be daring and try different odors for different occasions. You may be surprised at the results.

You'll need 1 teaspoon (4 g) of boric acid and 2 tablespoons (30 ml) of glycerin, into which you stir your favorite water-base perfume to suit your taste (smell). Then mix in 2 cups (474 ml) of water. When thoroughly mixed together, stir in 1 cup (237 ml) of isopropyl alcohol.* Use the mixture by splashing liberally on the face and neck after shaving.

(Sherman Toy Corporation, P.O. Box 455 WOB, West Orange, N.J., 07052 is but one of several companies that handle "essential" oils and perfumes.)

** See Appendix B for cautions pertaining to this ingredient.*

### *After-Shave Talc*

Using an after-shave talc serves two purposes. First, and most important, it reduces the shine which appears on the face following shaving; and second, the ingredients serve as an antiseptic since most products contain compounds for that purpose. This helps prevent nicks and scrapes from becoming infected and reduces the opportunity for pimples to start from surface bacteria.

If you use talc after shaving or have been thinking about buying some but just haven't gotten around to it, you may wish to mix your own special brand with the formula which follows. You'll find it just as satisfactory as the commercial products (perhaps even better), as you can use your own perfume or cologne for your favorite odor.

To make, mix 2 cups (454 g) of talc, 1 tablespoon (14 g) of boric acid, 5 teaspoons (20 g) of magnesium stearate, 1 tablespoon (14 g) of zinc oxide,* and an oil-based perfume — odor and quantity of your choice. A few drops should be sufficient, however, so don't overdo it.

### *Antichap Handcream*

Those who work with their hands can appreciate the advantage of having them smooth and soft, which prevents cracks, bleeding, and a lot of annoyance. If not attended to, hands which become chapped can develop serious problems with severe discomfort. Those who work outdoors and are exposed to the elements are especially prone to problems associated with skin chapping.

There are many good handcreams available on the market today which will go a long way in helping reduce the problem of chapped skin. For those whose activities subject them to the problem on a constant basis, the cost of the commercial products can become alarmingly high. If you find this to be the case, the following formula might be of value to you. Give it a try, and I think you'll end up using it in lieu of the store-bought products.

** See Appendix B for cautions pertaining to this ingredient.*

Mix 2/3 cup (158 ml) of glycerin with 1/3 cup (79 ml) of denatured* or isopropyl alcohol,* stir in 2 tablespoons (30 ml) of peanut oil, then add 1 teaspoon (4 g) of powdered tragacanth and 2 teaspoons (10 ml) tincture of benzoin (in that order). Rub into all exposed and affected areas of the skin, paying special attention to the knuckles, wrists, and backs of hands.

*Antichap Stick for Lips*

Lips serve a multitude of purposes and they deserve a little attention to function properly. This means protection from the drying sun in summer, as well as protection from the wind and cold of winter.

My first antichap stick was issued to me when I was in the Army. It was effective in protecting my lips from the weather of the constant outdoor life, and I was sorry I hadn't enlisted earlier (rather than waiting for the draft). It's surprising how little it took to make some GI's happy.

While a lot of years have passed since that first encounter with lip contentment, I find that a good antichap stick is as necessary today as it was back then. But instead of reenlisting, I make my own with a few simple ingredients and a few minutes' time. The finished product works just as well as the commercial types, and I can indulge myself in a few new wet trout flies with the money I save in doing so.

Melt ¼ cup (57 g) of beeswax (refined grade preferred but not necessary) in the top of a double boiler. When melted, add ¼ cup (59 ml) castor oil, 3 tablespoons (45 ml) of sesame oil, and 2 tablespoons (28 g) of anhydrous lanolin. Allow to partially cool and pour into small containers that have been sterilized.

I find the simplest way is to use my finger to apply the formula to my lips. Any that is left over goes to the back of my hands, which also enjoy the protection.

*Athletes Foot Ointment*

Most people know that a fungus is responsible for the condi-

** See Appendix B for cautions pertaining to this ingredient.*

tion known as athletes foot, and since the parasite needs moisture to develop, keeping one's feet dry is an aid in reducing infection. In spite of a lot of rubbing, and powder in my shoes, I always managed to come down with a good case of this dermal parasite each football season when I was in high school. In fact, the entire football squad had it — and I thought rubbing a towel between your toes to relieve itching was part of the normal warm-up before each practice and game.

Good sanitation in the lockerroom goes a long way in keeping the fungus from getting established. Washing floors and drains with a strong solution of bleach that contains chlorine after each use is effective. But when all else fails and you wind up with that itching feeling on your toes and feet, you'd better try one of the effective commercial products — or make a mixture of your own that can suppress the fungus before it spreads and causes severe discomfort.

Heat 1 cup (227 g) of anhydrous lanolin in a double boiler until it melts, then stir in 1 tablespoon (14 g) of flowers of sulfur, making sure they are well mixed. Cool and, before it solidifies, pour in jars for future use. Apply to infected and surrounding areas of the feet and toes to prevent secondary infection. Consult a physician in cases where severe cracking and bleeding occurs.

### *Brilliantine*

I'm not sure how many use a brilliantine-type hair dressing today, but there was a time when the slick-look was "in" and the dry-look definitely "out." And who knows — the way styles change, we're bound to run out of new ideas and go back to the old ones again. That certainly appears to be the case with my trousers this year, as I've noted the return of pleats in the front. If you decide to try the old part-down-the-middle with a slicker-look routine and can't find the proper commercial hair dressing, why not use the following formula and mix your own? It'll work just as well.

Mix 1 cup (240 ml) of glycerin into 1 cup (240 ml) of denatured alcohol.* Add a water-soluble perfume in quantity and fragrance to suit your preference.

* *See Appendix B for cautions pertaining to this ingredient.*

*Brushless Shaving Cream*

No matter how hard you try, it's difficult to beat the shave you get using shaving cream and one of the improved safety razors. I particularly like them as I'm able to remove dried skin while getting a close shave. This leaves my face cleaner, and I think it reduces the wrinkles that form particularly on the neck and chin.

There are obviously a lot of others who agree with me, judging by the number of products for this purpose that may be found on the shelves of our drugstores and markets. They come in creams, aerosols, and liquids, with fragrances that would have made even the famous dancehall girls of our Wild West blush. You might not smell as good, but you can get almost the same shaving results if you formulate the following product for your brushless shaving cream.

Heat to a simmer 2 cups (474 ml) of water, then stir in ½ cup (113 g) of stearic acid until melted. Remove from the heat and add ⅛ teaspoon (0.6 ml) of triethanolamine.* If you like, add a few drops of your favorite cologne or shaving lotion, to your specifications. Cool to room temperature and pour into jars. To use, just wet your face with hot water and rub the cream on your beard. You'll be amazed at how smooth a shave you'll get.

*Contact Lens Cleaning Fluid*

Keeping contact lenses clean is about as important as keeping up with them. I've often wished there was some good way of finding those small disks once they are removed. Short of labeling them with an isotope and using a Geiger counter. I haven't come up with anything to recommend. It's still hunt and seek, and without the aid of the lenses, one finds oneself at an immediate disadvantage. While I haven't a solution to this problem, I have a "solution" you can make that cleans them satisfactorily.

The commercial products are convenient and inexpensive; however, there are times when your supply is depleted or unavailable and you may wish to make your own. If and when that time comes, dissolve ⅛ teaspoon (0.5 g) of sodium bicarbonate

* *See Appendix B for cautions pertaining to this ingredient.*

*Denture Cleaner*

Keeping dentures clean is as important as keeping the natural teeth clean, not only for aesthetics but for good oral hygiene as well. Industry has used chemistry to the fullest extent in developing our modern-day dentures. They are rugged (compared to early models), natural looking, and quite effective. But regardless of their virtues, they require cleaning and daily cleaning is strongly recommended. When possible, the entire plate should be removed and soaked for at least 15 minutes in a good cleaner.

One cleanser you can formulate yourself for a few cents is made by mixing 1 teaspoon (4 g) of 10% citric acid into 1 cup (237 ml) of denatured alcohol* or isopropyl alcohol* and adding 10 drops (1.5 ml) of peppermint oil. After soaking for at least 15 minutes, rinse in clean water and you're ready to replace them, with assurance that your dentures are clean, odorless, and socially acceptable.

*Face-Powder Base*

A good face-powder base is important in making up the face. It serves as the beginning of operations, and the choice of shades of lipstick, rouge, or other tints depends a great deal on the base. Naturally, the color of the skin dictates the process and the selection of colors. And well-groomed individuals blend the colors of the makeup skillfully with the garments they intend to wear. We've all encountered someone who had used a shade of lipstick that was totally mismatched for the color scheme of her clothes, which detracts from a beautiful face and very fine clothing.

If you enjoy making things for yourself, as well as making yourself beautiful with the help of cosmetics, you may be interested in this formula for a satisfactory face-powder base. Simply mix 4½ cups (1 kg) of 5-grade talc with 1 cup (227 g) precipitated chalk and 1 cup (227 g) of zinc oxide.* Mix and sift. This produces a basic white face powder; it may be perfumed as you

* *See Appendix B for cautions pertaining to this ingredient.*

desire. Use sparingly at first, in case you're allergic to the compounds.

*Fingernail Bleach*

Those who work around chemicals and dyes find that their hands develop various shades of colors. The same goes for anyone involved with machinery and oil. While many cleansers do an effective job in cleaning the skin, most commercial products will not remove the stains on fingernails. These protective units on our fingers are principally calcium, which absorbs many oils and dyes, and about the only way to remove such unwanted stains is to use a bleach.

Formulating one is a simple process, and by soaking the fingernails in the solution for 10 to 15 minutes you can restore them to their original color. Simply mix ¼ cup (57 g) of sodium perborate in 3 cups (177 ml) of water until the chemical is completely dissolved. Label, and store in a safe place until you need it.

*Fingernail Polish Powder*

While the cosmetics industry has gone to a lot of trouble answering the color demands of those who use fingernail polish, there are many who prefer to polish their nails with a natural look rather than an exotic color. If you do a lot of rubbing to bring a luster to your nails, why not try a powder that will aid the polishing process?

To make your own, mix 4 tablespoons (56 g) of pumice powder, 1 tablespoon (14 g) of talc, and 5 tablespoons (70 g) of stannic oxide. To use, apply to nails with a damp cloth, polish, rinse with water, and wipe dry.

*Footbath Powder*

Have you ever thought how little care most of us take of our feet? Especially compared with the effort we spend dressing and pampering our hair? With hair, about all we derive from our efforts is the aesthetic effect, with little physical response derived from the money and labor involved. Oh, that we should devote even half as much attention to those unsung members that are responsible for supporting our full weight, cramped into leather shells and forced to move as we command! How many times have you treated yourself to a 10- or 15-minute footbath after a long and strenuous day? Not too frequently, I bet, and for that reason I'd like to suggest you begin to pamper your feet a little more. No good foot powder to use, or too expensive? No more procrastination, since the following formula will help eliminate poor excuses.

Mix ⅛ teaspoon (0.5 g) of menthol crystals, 4 tablespoons (56 g) of powdered alum, ½ cup (113 g) of boric acid, and 2/3 cup (151 g) magnesium sulfate. After mixing, use at the rate of 1 teaspoon per gallon of hot water.

Store the excess in a tightly sealed plastic jar for future use. Label well and be comfortable in the thought that you're indulging in an inexpensive luxury that is also healthy. Try it after a hard day's work and you'll begin to wonder why you haven't done it before.

*Greaseless Hair Dressing*

Here's a good hair dressing, especially for the outdoor person who finds the sun and wind taking their toll on hair and does not like a type of dressing that leaves the hair greasy. I find myself collecting a lot of dirt and grime when I've used the greasy type, but the following formula is satisfactory and economical to make. If you're about to reach the bottom of your hair dressing container, why not get the ingredients and try them before investing in another jar or tube of the commercial type?

Stir 1 cup (237 ml) of glycerin into 1 cup (237 ml) of isopropyl alcohol.* Then mix in a water-base perfume to suit your odor preference; you can use your favorite cologne if you like. Use as desired to keep the hair well groomed.

* *See Appendix B for cautions pertaining to this ingredient.*

*Hair Cleaner*

Look on the shelves of any drugstore or supermarket where personal-care products are found and you'll be amazed at the space devoted to compounds for the maintenance of one's hair. There are endless brands of conditioners, dyes, tints, curling compounds, and shampoos — almost as many as there are heads to be washed. We spend nearly as much money on our hair as for all other products we use in personal care. Longer hair styles account for a large percentage of this expenditure. Many barbershops have combined with beauty salons where "hair stylists" cater to both sexes, "styling" hair for $10 (or more) that they used to trim for $1.50. Nevertheless, well-groomed hair goes a very long way in improving personal appearance.

Those who still haven't reached the decision as to the best shampoo for their particular hair condition might wish to try the following hair cleaner. It's a paste-type shampoo, which relieves that unpleasant habit of spilling when you're in the shower or bending over the sink with suds in your eyes.

To 1½ cups (341 g) of ammonium stearate add approximately ½ cup (118 ml) of water. Include a water-base perfume in the odor and strength you desire. Vary the amount of water to obtain the consistency you favor. If your water is ladened with minerals, it's a good idea to use rain water or distilled water.

Try this formula. I think you'll like it, as it leaves your hair squeaky clean.

*Hair Shampoo*

I've always been a shower man myself, and that means a daily shampoo. As a consequence, I go through a lot of shampoo and soap. Because of this, I've had the opportunity of trying just about every brand name that has come on the market over the past couple decades. While I find most of the commercial products satisfactory, I'm just as satisfied with one that I can make at home for less than half the cost of those from the store. Besides, by using my favorite cologne in preparing it, I have the fragrance of my choice and not one that leaves me smelling too feminine.

To make your own hair shampoo, stir 1¼ cups (296 ml) of

oleic acid, 1 cup (237 ml) of coconut oil, and 1¼ cups (296 ml) of triethanolamine* with a water-base perfume or cologne to suit your individual olfactory system. This mixture is a concentrate and should be used with enough water to obtain the lather consistency you prefer.

### *Hand Lotion*

Our modern communication network has made us painfully aware of how important it is to have hands you love to see, love to touch, and — according to some TV ads — also love to smell. It's beyond me how our pioneers ever survived the rigors of life without knowing the importance of lovely hands. Or did they? There was probably just as much interest back then as today, but the reasons for keeping the hands soft and pliable were probably for practical rather than aesthetic reasons. One very popular commercial hand lotion that is sold today is, with a few modifications, the same as the lotion used by my grandmother almost 100 years ago.

Certainly the advantages of well-groomed hands are obvious. We all know dry hands are not only hard to keep clean but may chap and become painful. To avoid this, application of a good lotion after washing your hands and before going into inclement weather is a smart maneuver.

Since daily use will deplete a large bottle rather rapidly and the cost can add up, here is a satisfactory hand lotion formula that will do just about everything a commercially prepared product will do. It costs less than half the price when you make your own. You can also make up your own name for its psychological effect, such as "Buttercup Creme" or "Lavender Lotion" (whatever turns you on).

Heat 6 tablespoons (90 ml) of glycerin, the same quantity of anhydrous lanolin, and ½ cup (113 g) of petrolatum in a double boiler until all are liquified. Then mix in 1 tablespoon (14 g) of boric acid and a few drops of any oil-base perfume which pleases you. Cool and place in a bottle, where it can be stored until used as a soothing lotion for dry and chapped hands.

* *See Appendix B for cautions pertaining to this ingredient.*

*Ink Remover for Hands*

I recently visited a friend of mine who proudly displayed an unfinished "antique" school desk like the ones I used many years ago. In one small rural school I attended, we had a double seat where two of us sat at each desk. I'm sorry I hadn't thought of it over the years, as I could have used it as an excuse for my neglected education. It was hard enough, back in those days, to keep your mind on studying, but with someone *that* close to you to aid in distractions, it was even more difficult to accomplish anything worthwhile. In each desk was an inkwell, which was always a source of problems. Hands — and most of the desk — were always discolored as we practiced penmanship. What I would have given for a good ballpoint pen back in those days! And I remember how frequently my fountain pen leaked and the entire shirt pocket, plus inner garments, became a brilliant blue as the ink spread. Oh, well, we survived, in spite of the frustrations, and today find them amusing as we reminisce about them.

Many of our everyday activities still bring us into contact with ink through various machines, pads, and marking instruments. When we end up with our fingers smeared with various-color compounds, a simple formula can be mixed which works well in removing most ink products.

In 3 tablespoons (45 ml) isopropyl alcohol,* add 1 tablespoon (15 ml) of glycerin and 1 tablespoon (15 ml) of hydrogen peroxide. Rub a small amount into the stained area and wash with clear water. Repeat as necessary. Most inks can be removed quite easily with this treatment.

*Lime and Glycerin Cream*

A delightful and effective moisturizing cream can be formulated with simple ingredients and a few minutes' time. Those who are exposed to wind and sunshine realize how important it is to use some type of cream or lotion that prevents the skin from becoming excessively dry. As we all know, when this happens wrinkles begin to appear and we begin to worry, and worry can induce ulcers, and ulcers can cause poor health. So it's much

* *See Appendix B for cautions pertaining to this ingredient.*

simpler to use a cream that eliminates all these problems — or at least tends to reduce them.

In your kitchen or bathroom, mix ⅛ teaspoon (0.5 ml) of glycerin and 1 cup (237 ml) of lime water (squeeze a fresh lime); then stir in ⅛ teaspoon (0.5 ml) of almond oil* and ⅛ teaspoon (0.5 ml) of lemon oil. To use, pour a small amount into your palms and rub lightly on your face and neck. The effect is both fragrant and refreshing.

Since there are no preservatives in this formula, mix small quantities, or keep it sealed and refrigerated.

### *Makeup Remover*

I'd hate to estimate how many pounds of makeup are used daily, but it must run into several tons. And what goes on must come off; so a lot of makeup removers are used each day as well. Look over the cosmetics shelves and you immediately get the idea that this line of personal products is big business. There is a color, odor, and bottle size for the most discriminating, with prices to match.

If you're one who uses large quantities of makeup remover and seems to be constantly exhausting your supply, why not try formulating one that is just as effective, much cheaper, and really simple to put together?

Place ½ cup (113 g) of beeswax, ¼ cup (57 g) of stearic acid, and ½ cup (118 ml) mineral oil in the top of a double boiler. After the ingredients have melted and blended, remove them from the heat, cool, and pour into jars. Apply with tissue to remove lipstick and other makeup.

Since some people are allergic to different compounds, it would be good to try this on a small area of the skin to ensure there would not be a reaction.

### *Mouthwash Powder*

There's got to be a mouthwash for just about every like and dislike that mankind (personkind?) can imagine. The shelves are

* *See Appendix B for cautions pertaining to this ingredient.*

stocked with about every color and fragrance that can be combined. And most are effective (to a degree) in helping keep the mouth clean, thereby reducing odor buildup.

If you are a mouthwash addict (and I'm certainly one), you may wish to concoct your next mouthwash powder and rate it against your present brand (which you can call "Y"). I think you'll find it just as pleasant and effective — and a lot cheaper — as what you're presently using.

Add 2 tablespoons (28 g) of cornstarch to 1 cup (227 g) of sodium perborate. Then mix in 15 drops of peppermint oil. Add 2 tablespoons to a cup of warm water, stir, and gargle. It leaves you kissing sweet.

*Nourishing Cream*

I haven't the slightest idea who coined the name "nourishing cream" for cosmetics that are used as moisturing agents. Perhaps the idea developed more in the spirit of sales promotion than from a "nourishing" effect on the skin. Whatever the reason, the long line of products that are available for the purpose of reducing dry skin and wrinkles testifies to the effectiveness of such creams. Certainly, by using them frequently one can detect a decided advantage.

If you use cosmetics that prevent your flesh from burning off the bone, I suggest the following formula for a "nourishing cream" that should provide as much absorption as the commercially prepared products you've used in the past.

To concoct this cosmetic cream, put 2 tablespoons (28 g) of anhydrous lanolin, ¾ cup (10 g) of stearic acid, and ½ teaspoon (2.5 ml) of triethanolamine* in the top of a double boiler. Heat until melted, then stir in 1 cup (237 ml) of water and sufficient water-base perfume. Apply by rubbing into your hands and face, elbows, and other exposed areas of the skin each evening before going to bed. Remove it in the morning by wiping or washing off.

* *See Appendix B for cautions pertaining to this ingredient.*

*Softening Bath Water*

Water in most areas of the United States is "hard" from the various minerals in it. In the Southwest, sodium and calcium salts are common, and in some areas it's downright difficult to get soap to foam or bubble at all. The same may be true for areas with high concentrations of iron. A good bath under such conditions is a little difficult without the aid of a water softener. Unless you resort to one, you sometimes come out of a bath feeling as though you're dirtier than when you started.

A simple solution to the problem is adding 2 tablespoons (28 g) of sodium sesquicarbonate to the bath water. You may need more, depending upon the quantity of water you use or the type and amount of minerals in your area.

*Sore Muscle Rub*

As we put our out-of-shape bodies back into shape, we frequently find forgotten muscles protesting the sudden demands placed on them. And as we grow older, we find ourselves more prone to sore muscles with very little provocation. Every year, before the opening of the hunting season, I promise myself that I will get my old body into shape, but while my intentions are good, my actions aren't worth much. Like many others, I drag myself into bed after the first few days with aching muscles that are just crying for attention. If you're home, where you can take a good hot bath, you have the advantage, but invariably I'm on top of a mountain or in a canyon that's miles from any hot water. In such cases, I resort to rubbing the tired and aching muscles with a formula that eases the agony of getting up the next morning. I mix a fresh batch before the season and see that the supply remains on hand in the first-aid box. I've also found that out-of-shape dogs don't mind a little attention with the same material at the end of the day. I keep them from licking it by wrapping a towel or blanket around them.

This sore muscle rub is made by dissolving ½ cup (118 ml) of methyl salicylate in ⅛ teaspoon (0.5 g) of camphor, ⅛ teaspoon (0.5 g) of menthol, and 1½ cups (355 ml) of isopropyl alcohol.* When thoroughly blended, mix in ¼ cup (59 ml) of mineral oil. To use, pour out a small quantity on the hand or directly on the area of the body to be massaged, and rub in, massaging the sore muscles.

### *Tooth Powder*

Oral hygiene has come a long way in the past 50 years. We know much about the proper care of our teeth and the reasons behind the problems that arise by neglecting them. It's common knowledge that cavities (or caries, as they are more correctly referred to by the dental profession) are principally the result of food particles — especially those with sugar — adhering to the teeth. Various dentrifices are available that are highly effective in removing such particles. By "good brushing," aided by a dental thread or "floss," you can reduce the chances of developing "problem teeth." (I'm sure you hear this each time you visit your dentist; it's so simple but so often neglected.) Many individuals find common table salt alone, or in combination with baking soda, is all they need for cleaning their teeth. If you're one of these, but would like to use a slightly more pleasant as well as effective dentifrice, try the following recipe.

Mix ¾ cup (170 g) of sodium chloride, ¼ cup (57 g) of sodium bicarbonate, 2 tablespoons (28 g) of magnesium bicarbonate, and 1 tablespoon (14 g) of sodium perborate — together with 15 drops (1.5 ml) of peppermint oil. This will produce a formula that you should find acceptable but inexpensive. My guess is you'll never go back to salt and soda again.

### *Vapor Inhalant*

Modern medicine is still fighting that old enemy of man, the common cold. It accounts for more lost days of work and general misery than all other ailments. Since antibiotics are generally of

* *See Appendix B for cautions pertaining to this ingredient.*

little value, the old remedy, "Take an aspirin and get plenty of rest," still seems about as good as anything. But when nasal passages become stuffy, throat sore, and lungs congested, I find it's time for more than rest and an aspirin.

Years ago, one of the most common remedies for colds, the flu, and even serious cases of pneumonia was a supply of vapor (usually from a boiling kettle) to which pine rosin had been added. Many of us are alive today because of this old-fashioned cure. We might have survived without it, but it usually wasn't long before I was up and well again after several nights' sleeping in an area saturated with such vapor.

If you haven't added this procedure to your repertoire of common head-cold remedies, you might try it the next time you're plagued with the problem. Instead of a kettle, you can use a safe, commercial vaporizer that can be set in the area where the patient rests or sleeps. (If you have respiratory problems, consult a physician before using it.)

I like to use an inhalant that increases the effectiveness of the vapor. One that is effective, inexpensive, and easy to make, involves mixing ¼ cup (59 ml) of pine needle oil and ¼ cup (59 ml) of tincture of benzoin. Use 1 teaspoon (5 ml) in 1 quart (1 l) of water, placed directly into the vaporizer. Also, pine needle oil can be extracted by steam distillation of pine needles, using a double boiler. It's simple, and just as effective as oil purchased from the store.

Try the remedy the next time you have trouble; you'll be glad you did!

## NOTES

Name of Formula: ______________________________

Date Made: ____________________

Ingredients and amounts: ______________________

______________________________________________

______________________________________________

______________________________________________

Label: Ingredients and caution warnings

Observations: _________________________________

______________________________________________

______________________________________________

______________________________________________

______________________________________________

- - - - - - - - - -

Name of Formula: ______________________________

Date Made: ____________________

Ingredients and amounts: ______________________

______________________________________________

______________________________________________

______________________________________________

Label: Ingredients and caution warnings

Observations: _________________________________

______________________________________________

______________________________________________

______________________________________________

______________________________________________

NOTES

Name of Formula: ______________________________
Date Made: __________________
Ingredients and amounts: _________________________
______________________________________________
______________________________________________
______________________________________________

Label: Ingredients and caution warnings
Observations: _________________________________
______________________________________________
______________________________________________
______________________________________________
______________________________________________

- - - - - - - - - -

Name of Formula: ______________________________
Date Made: __________________
Ingredients and amounts: _________________________
______________________________________________
______________________________________________
______________________________________________

Label: Ingredients and caution warnings
Observations: _________________________________
______________________________________________
______________________________________________
______________________________________________
______________________________________________

# 3

# Automotive and Workshop

## Automotive and Workshop

Caulking Compound
Concrete Cleaner
Concrete Waterproofing
Fiberglass Filter Treatment
Mechanics' Hand Soap
Metal Degreasing Solution
Metal Polish, All Purpose
Outdoor Wood Treatment
Paint and Varnish Remover
Penetrating Oil
Radiator Cleaner
Radiator Scale Prevention
Rust Remover for Tools
Sculpture Modeling Wax
Undercoat for Shellac
Whitewash for Adobe and Stucco

Many veterans (of several wars) who served in the Navy had firsthand experience with a chemical reaction called corrosion. Perhaps they knew it better as rust, and well remember the long hours, weeks, months, and even years they spent chipping away at metal surfaces to remove old layers of paint and rust before adding new paint. Though it may not have been much consolation at the time, it would have relieved the boredom of the "chip and paint routine" had a down-to-earth explanation been given for the process of corrosion, with rust as a final product.

While the process is fairly complicated, it is an electrochemical reaction or exchange of electrical current between two areas. These areas are known as the anode and the cathode, and they gain and lose electrical charges. When metals are exposed to heat, water, or air, the electrical charges are speeded up and the compounds attempt to revert to their natural state. Steel, upon being heated, forms scale or oxides, while iron, when exposed to water or damp air, forms hydrous oxide (better known as rust). Thus the metals revert to the oxide form they were in when removed from the ore. This is caused by the electrical charges of the metals, the cations and electrons moving outward as they are attracted to the oxygen of the atmosphere.

Why do metals differ in their ability to withstand corrosion? Since gold is heavier and denser than iron and its atoms are packed into its molecules differently, there is greater resistance to the process of oxidation or loss of electrical charges. As the initial film of oxides is formed, the corrosion process is slowed, but as time passes, the rust is porous to the oxygen and water, and corrosion increases, until finally the entire metal is consumed. We add metals with low electrical conductivity to other metals to form alloys, thereby reducing the oxidation process. When (for various reasons) this isn't possible, we go back to the old process we know as painting. By the application of various metallic-base paints we actually prohibit or retain the electrical charges from coming into contact with oxygen. Anticorrosive pigments are metallic coatings that bind with the metal to prevent contact with the environment and subsequent corrosion. In time, these coatings are broken down by microbial action (fungi and bacteria) or chemical and physical reactions with the atmosphere, and must be replaced.

Somewhat the same process of protection against corrosion

occurs when we apply oils or grease to metals. As long as the film remains on the surface, protection from oxidation occurs and scale or rust does not develop.

While you may not feel much better about spending a lot of time cleaning and painting metal surfaces to protect against corrosion, you may keep yourself amused by remembering the interesting chemical changes that are attempting to keep one step ahead of your efforts. Good luck — you'll need it!

### *Caulking Compound*

That leak around the kitchen sink or the wind that blows in around the windows on winter days as the result of caulking compound becoming brittle and falling out are annoying to the homeowner. They may be simple to fix, but getting around to remembering to buy the caulking compound is another matter.

Instead of remembering to buy a commercial compound, use this formula. Make a note to buy the ingredients at the hardware or paint store and formulate your own caulking compound in the quantity needed for the particular job on hand. It's fun, less costly, and the next time you need it, you're not faced with the problem that often occurs: finding that the remainder in the container has become dry and hard.

To 1¼ cups of kerosene* (296 ml) add 1½ cups (340 g) of asphalt, dissolve, and then mix in 2 cups (480 g) of powdered asbestos* to a puttylike consistency. (WARNING: Do *not* breathe asbestos dust; wear a protective mask when mixing asbestos.) Use around windows, tubs, showers, and sinks as needed.

### *Concrete Cleaner*

In spite of all efforts to keep oil and rust stains from our garage floor and driveway, we somehow, sooner or later, find ourselves faced with the chore of cleaning such spots.

Some stains are persistent, but a formula that I've found to be effective as a concrete cleaner may be mixed quite simply in a few minutes' time. When the concrete is washed, diluting the solu-

* *See Appendix B for cautions pertaining to this ingredient.*

tion with water, it will not harm grass or other plants. Some of the commercial compounds can be more effective, but there are times when they are toxic and hazardous to vegetation, which may limit their use.

If you find yourself in such a situation, use gloves to mix 3¼ cups (738 g) of sodium metasilicate into ¾ cup (170 g) of trisodium phosphate* and ½ cup (113 g) of soda ash. Stir with a wooden spoon or stick until thoroughly blended. To use, wet the concrete thoroughly, then sprinkle the mixture over the spots. Allow to stand at least 15 minutes, then rinse well with clean water. Repeat as needed.

Some rust stains may be difficult to eliminate due to the porosity of the concrete and the large quantities of iron in local water supplies.

### *Concrete Waterproofing*

Those who have basements in areas with high water tables are constantly confronted with the problem of moisture penetrating through concrete walls and floors. The increased humidity causes a musty smell to develop—the result of molds' becoming established on the wet surfaces. Besides the unpleasant odors (which can penetrate into the upper portions of the house), stored foods and other items may fall victims to fungi. If the moisture condition continues, there may be structural damage to the building, which can result in expensive repair.

In this case, 1 ounce of prevention is worth more than 100 times the cost of the cure, and calls for every effort to waterproof the permeable concrete. Modern chemistry has done much to formulate compounds that are effective waterproofing materials, and they may be found on the shelves of every good hardware and paint store. Unfortunately, they were not available during the construction of many of our earlier dwellings, and the result is that moisture penetrates the concrete surfaces.

For those who may not have access to modern waterproofing compounds, there is a formula which can be very easily concocted, is highly effective, and costs substantially less than purchase of the manufactured products.

** See Appendix B for cautions pertaining to this ingredient.*

To prepare, mix 2½ pounds (1.1 kg) of ammonium stearate in 8 gallons (32 l) of water. Stir until dissolved and apply 2 or 3 coats with a masonry brush to the concrete surfaces. Store any leftover in plainly labeled plastic bottles for future use. Best results can be expected if the formula is applied when surfaces are dry.

### *Fiberglass Filter Treatment*

The use of fiberglass in filters for furnace and refrigeration units to remove dust particles from the air has increased their efficiency tremendously. There is less resistance to the stream of air and the undesirable foreign particles are retained in greater percentage than by many other materials used for the same purpose. Where dust is extremely heavy, it may be desirable to increase filter efficiency by treatment with a compound that can be formulated inexpensively and with a few minutes' time.

To 5 cups (1.2 l) of paraffin oil add 2 cups (554 g) of stearic acid and 1½ cups (340 g) of bentonite, stirring well. Separately, mix ¾ cup of triethanolamine* with 2½ gallons (10 l) of water, then combine this with the first mixture, stirring until a cloudy emulsion forms. To use, simply spray on the filter (after cleaning it) or on an untreated fiberglass filter. Don't forget to clean frequently, as dust will accumulate more rapidly on treated filters.

You may wish to make your own cleaning compound by adding 1 teaspoon (4 g) of sodium metasilicate to 1 teaspoon (5 ml) of liquid detergent and mixing this with 1 quart (1 l) of water. Brush this onto a wet filter and hose with a strong stream of water. Treat again before reinstalling.

### *Mechanics' Hand Soap*

I've never fully enjoyed working on engines or automobiles because, number 1, I don't always know what I'm doing and, number 2, I end up with more dirt and grease on me than the

** See Appendix B for cautions pertaining to this ingredient.*

piece of equipment I'm repairing. Even 1 week later, my hands look as though I have forgotten to wash them during the past seven years. Those of you who have the ability, skill, and time to perform your mechanical needs and find, after completing the task, that you're in need of a good hand cleanser, might try the one I've found to be as effective as the commercial types. I don't consider myself an expert mechanic, but after using this formulation I've got the cleanest hands on the block. Mix some of your own and I'm sure you'll be happy with the results.

Mix 2¾ cups (651 ml) of denatured* or isopropyl alcohol* in ½ cup (118 ml) of Stoddards solvent.* Slowly stir this mixture into 2 cups (474 ml) of coconut oil, then add enough lemon oil for the fragrance you prefer. Pour a small amount into your hands, rub well, and rinse with water or wipe with paper towels.

### *Metal Degreasing Solution*

The accumulation of grease on motors and engines may raise their temperature high enough to be injurious to the normal operational process. In time, this can result in permanent damage, necessitating replacement of parts or, worse, a new engine. It also creates a fire hazard. Aviation, following World War II, produced excellent degreasing compounds that are now commercially available (many in aerosol containers which are handy to use). If you're inclined to keep an engine — or any metal surface, for that matter — free of excess oil and grease but find the commercial products unavailable, you may wish to try formulating one of your own. It is easy to make, easy to use, and easy on the pocketbook. It's a good material to use as the initial application to eliminate most of the problem. Then give the surface a light second coat with a commercial product that can be washed off with water from a pressure hose. Applied in this manner, you can eliminate excessive wiping.

Simply mix 1 quart (1 l) of triethanolamine* with 4 tablespoons (60 ml) of boiled linseed oil. Apply it to the grease-coated surface, allow it to soak in for 5 to 10 minutes, and then wipe clean.

* *See Appendix B for cautions pertaining to this ingredient.*

*Metal Polish, All Purpose*

Tarnished metal objects can be polished back to their original shine if foreign particles have first been removed with an appropriate metal cleaner. This can be accomplished with minimum effort, both in formulating a polish from very simple ingredients and in using the finished product. I've found it quite effective on the chrome surfaces of my car and boat. It also does an excellent job in polishing hubcaps that have been subjected to road-dirt build-up.

Carefully mix ½ cup (118 ml) of household ammonia with ½ cup (118 ml) isopropyl alcohol,* then stir in 1 cup (227 g) diatomaceous earth in small amounts at a time. Add just enough water (½ cup [118 ml] or less) to achieve a thick, creamy consistency. Shake well before each use, apply with a sponge or cloth and rinse with clean water.

*Outdoor Wood Treatment*

The demand for a good wood treatment has increased as the number of summer homes and cabins built with unseasoned lumber or peeled logs has almost doubled in the last decade. Because these materials are frequently subject to severe weathering, it's a good idea to treat unpainted wood to help prevent its drying and cracking, as well as assist in repelling moisture.

An inexpensive wood treatment that has proved highly effective for exposed wood that isn't painted is a mixture of 1 quart (1 l) of boiled linseed oil and 1 quart (1 l) of turpentine.* Use a regular paint brush to apply as often as needed.

*Paint and Varnish Remover*

Those who refurbish old homes or refinish antique furniture know what it means to have a good paint and varnish remover available for the job. One of the hardest parts of the process can be removal of a half-dozen layers of paint that may date back to the 1800s. If you find yourself using large quantities of remover and

* *See Appendix B for cautions pertaining to this ingredient.*

the cost going up each time you buy it, the following formula may interest you. Give it a try, and if it isn't as satisfactory (and much less expensive) as your present paint remover, you can always go back to the old one.

Stir 1 cup (227 g) of caustic soda,* ¾ cup (170 g) caustic potash,* 2 cups (480 g) of calcium carbonate, and 1½ cups (341 g) of pumice powder together. To use, take ½ cup or so of this powder mixture and add just enough water to form the consistency of cream. Brush on and scrape off. WARNING: This mixture is highly caustic and can cause burns. Use gloves and if it comes in contact with skin, flush the area with water.

### *Penetrating Oil*

Several penetrating oils have come on the market in the past few years that are highly effective. Everyone has had the problem of trying to get a nut off a rusty bolt or removing a screw that was installed 50-odd years ago. No matter how hard we try, some are "frozen," and without special help we're bound to tear the threads on the bolt or the heads off the screws. However, a shot of penetrating oil and a few hours' time are generally all one needs to resolve the problem.

Those who enjoy making their own products and saving money while doing it can make an effective penetrating oil by mixing 1 tablespoon (15 ml) of butyl alcohol and 2 tablespoons (30 ml) of kerosene* into 7 tablespoons (105 ml) of mineral oil. Apply on rusted or frozen hinges, bolts, or screws and allow to set 5 minutes or so before working them free.

### *Radiator Cleaner*

Manufacturers of antifreeze and coolants for motor vehicle radiators recommend that, after varying periods of time, the materials be drained, the system cleaned and flushed and new antifreeze be placed in the radiator. Have you given serious thought to this recommendation? How long has it been since you've drained the cooling system on "old Susie," or that new model for that matter?

** See Appendix B for cautions pertaining to this ingredient.*

If this jogs you into action, you might wish to save yourself a trip to the auto dealer or filling station, as well as save yourself money, by borrowing some from the laundry room.

A simple and effective radiator cleaner can be made by adding 2½ pounds (1.1 kilo) of washing soda to 3 quarts (3 l) of water. Stir the soda into the water and add the mixture to the drained radiator, filling the rest of the way with water. After the engine has run for 30 minutes, drain again, flush with clean water for 5 minutes, and refill with your antifreeze-coolant compound. (If rust is present, a radiator rust remover may be needed.)

### *Radiator Scale Prevention*

In spite of all our efforts to protect the cooling system of our vehicles, adding water from various sources (which frequently have quantities of iron or calcium) can cause the formation of scale in the radiator. As the water and antifreeze heat and cool, the scaling process increases. In time, its build-up can clog the outlets in the system, and inadequate circulation results in overheating.

If you have an adequate supply of rain water for your radiator, you're lucky, since this kind of soft water assists in reducing scale and rust formation. If, however, you don't, you can take other measures that are easier than collecting rain drops. A simple scale preventor may be made by mixing 6 tablespoons (90 ml) of sodium silicate* and 2 tablespoons (28 g) of trisodium phosphate* with 1 gallon (4 l) of water. Put the mixture in the radiator and fill with water and/or antifreeze.

This precaution will pay dividends if you plan to keep that expensive piece of machinery for several years.

### *Rust Remover for Tools*

Remember that sinking feeling you had when you found your best crescent wrench rusting in the grass? You immediately promised yourself to take remedial action: first, a firm lecture to

* *See Appendix B for cautions pertaining to this ingredient.*

the children that no more tools would be misused (until next time) and, second, you'll get a rust remover, clean and oil the wrench, and have it ready when it's next needed. Sound familiar? It's happened to me with practically all the tools I've owned. My lectures on the proper use of tools, restricting their use, and locking them up seem to have proved useless, as have my numerous attempts to remember to buy a commercial rust remover from the hardware store on my next trip.

While I've never found the answer to the problem of the children's negligence, I have a solution to the second problem in an easy-to-make rust-remover formula for tools that need it.

Mix 1 tablespoon (14 g) of ammonium citrate crystals into 2 cups (474 ml) of water. Stir the crystals into the water until dissolved. Place the rusty tools in a plastic container (a gallon bleach bottle with the top removed is good) for 12 to 24 hours, remove, wipe, rinse in clean water, dry, and apply a light coat of oil or an antirust tool coating. Dr. Spock may have a better solution to your first problem.

### *Sculpture Modeling Wax*

Modeling wax can become an economic burden on the artist who uses great quantities in sculpturing. Commercially prepared compounds may not be sufficiently pliable, or some types may harden before they can be used. A good way to reduce its cost and to have modeling wax in the desired consistency is to formulate your own by following this simple formula.

In a double boiler (never melt flammable wax directly over heat), melt 2 cups (454 g) of beeswax with 4½ cups (1.6 l) of venice turpentine. When completely melted, add 2 cups (454 g) of lard (the commercial type is satisfactory) while the first mixture is still warm. Then add 1¾ cups (398 g) of powdered clay or talc.

Depending upon the consistency desired, the amount of clay or talc may be varied. Practice by making small quantities until you find the stability you determine to be most satisfactory for your needs.

*Undercoat for Shellac*

The demand for shellac is diminishing for use in finishing and protecting everything from photographs to the woodwork on TV sets, but before the advent of synthetic polyurethanes and similar products for wood finishes, shellac enjoyed much wider use. Today, much of our furniture is made from soft woods and there is greater need for a harder finish to protect the delicate surfaces. Still there are many uses for good-quality shellac finishes. Many antique refinishers still use it as their mainstay in obtaining that final luster.

To obtain that professional look when using shellac, try using an undercoat that will better fill the pores prior to the finishing coat — a simple and inexpensive process. Mix ½ teaspoon (2 g) of boric acid and 2 tablespoons (30 ml) of shellac, apply with a brush, and allow to dry to a hard surface before adding a second coat of shellac or varnish.

*Whitewash for Adobe and Stucco*

Tom Sawyer taught us how to manipulate a friend into whitewashing a picket fence, but he didn't give us the formula for the whitewash. Since that old and enjoyable story was written, we seem to have lost interest in painting the fence around the house or barn, or the wooden fences that surround the fields, with the lime mixture. Remember how clean and neat the farms used to look?

Interest in whitewash hasn't decreased, however, in areas where stucco houses and patio walls are principal components. This is especially true in the Southwest where adobe or cement brick may be surface-finished with cement stucco. As time and nature take their toll on these surfaces, it becomes necessary to refinish them to bring back a natural freshness. Whitewash can be just as effective as expensive paint, and may last just as long if the following formula is followed. Those with wooden fences can use it just as effectively.

Into 64 parts of water add 5 parts of adhesive cement (Wilhold or Blue Bird type), 4 parts of cement latex, and mix well with your hand or a paddle. When the latex is thoroughly mixed in, add 15 parts of Portland cement, adding slowly and stirring to avoid lumps or caking. Then carefully add 10 to 20 parts of hydrated lime, the quantity depending upon the whiteness desired. Lower rates produce a gray-white while higher rates of lime will whiten the final color. It's advisable to paint the lower-rate mixture on a small surface to get a good idea of the color after drying. Cements vary in color within geographic areas, so you'll have to do a little experimenting with the lime.

For those who wish colors, cement dyes may be added. Use the same precaution as for adding lime in determining the quantities to be added: try a small area and let it dry before doing the entire area.

It should be remembered that white reflects heat, making a home cooler in summer, which aids in conservation of energy. The ingredients for the formula are available through hardware, lumber, and paint stores.

## NOTES

Name of Formula: ______________________________

Date Made: ____________________

Ingredients and amounts: ______________________

______________________________________________

______________________________________________

______________________________________________

Label: Ingredients and caution warnings

Observations: _________________________________

______________________________________________

______________________________________________

______________________________________________

______________________________________________

- - - - - - - - - - -

Name of Formula: ______________________________

Date Made: ____________________

Ingredients and amounts: ______________________

______________________________________________

______________________________________________

______________________________________________

Label: Ingredients and caution warnings

Observations: _________________________________

______________________________________________

______________________________________________

______________________________________________

______________________________________________

## NOTES

Name of Formula: ______________________________

Date Made: ____________________

Ingredients and amounts: ______________________

______________________________________________

______________________________________________

______________________________________________

Label: Ingredients and caution warnings

Observations: _________________________________

______________________________________________

______________________________________________

______________________________________________

______________________________________________

- - - - - - - - - - -

Name of Formula: ______________________________

Date Made: ____________________

Ingredients and amounts: ______________________

______________________________________________

______________________________________________

______________________________________________

Label: Ingredients and caution warnings

Observations: _________________________________

______________________________________________

______________________________________________

______________________________________________

______________________________________________

# 4

# Animal Care and Agriculture

## Animal Care and Agriculture

Animal Dry Cleaning Powder
Animal Shampoo
Cattle Ringworm Control
Cattle Scabies
Dehorning Salve
Dog Foot Conditioner
Earwax Remover for Pets
Manure Fertilizer Activator

## Animal Care and Agriculture

The number of formulas in this section has been intentionally reduced because of available information from either the Cooperative Extension Service of your land-grant university or your local veterinarian. They have current information based on research generated by personnel from the College of Agriculture and the U.S. Department of Agriculture. Too frequently, this excellent source of information which is free or for a nominal fee for printed matter, is overlooked, especially by the urban population who is unfamiliar with the informational branch of our land-grant university system. If you have a problem concerning animal care, gardening, or general agricultural topics, look up your local Cooperative Extension Service in the phone book. It is usually listed under the county government or the name of the state's land-grant university. Problems which are beyond the scope of this group are referred to the research staff of the university, where they may begin research activities to seek answers that are not available.

Another reason for fewer formulas in this section pertains to federal and state regulations that control recommendations of many of the pesticides used in agriculture. The Environmental Protection Agency's regulations, provided in the Federal Insecticide, Fungicide and Rodenticide Act as amended, require that pesticide applicators of certain types of chemicals be trained and licensed through examination. Similar state regulations exist and most require that individuals who make recommendations for the use of pesticides also be licensed after study and examination. Such regulations have served to benefit some aspects of the environment; however, costs of farming operations have increased because of the costs involved in carrying out the programs and frequently less effective pest control practices. The public who has demanded more governmental action be taken to better protect the environment must be willing to support the EPA and its regulations through increased costs of food and fiber. Unfortunately, present regulations do little to control the sales and use of pesticides in the home. Over half of these chemicals are used for domestic purposes where they are stored and applied with no restrictions. The majority of illness attributed to pesticides is from this source and not from agricultural usage. Hopefully, better control of pesticides for domestic use will be forthcoming in the not too distant future.

## *Animal Dry Cleaning Powder*

Getting Old Bowser or Kitty Poo into the bath water may be a chore with some pets. It may also subject them to a case of distemper if they become chilled during cold weather. Both problems may be easily avoided by employing an animal dry cleaning powder that is simple to make and simple to use. You and your pet will probably enjoy not having to go through the usual wash-day hassle.

Mix 2 tablespoons (28 g) of trisodium phosphate,* 2 tablespoons (28 g) of Borax, 5 tablespoons (70 g) of sodium carbonate, with 14 tablespoons (200 g) of talc, into 1½ cups (340 g) of starch. Blend together with a fork and store in a dry plastic container until you're ready to use it. Rub in against the direction of the hair, and brush vigorously or vacuum to remove. A drop or two of pine oil or your favorite cologne may be added if you have a desire for a slightly different odor than is normally associated with dogs or cats.

## *Animal Shampoo*

Never in our history have we had more pets than today. They range from the standard to the bizarre, but regardless of what they are or what they look like, there is strong evidence that we are exhibiting our loss of contact with nature through the selection of companion animals. We shower love and care on them and they respond with comfort and companionship. But, humans, there comes a time when even friends must part — unless we take care of undesirable body odors with a good bath.

While a lot of good animal shampoos can be found in everything from pet shops to the local drugstore, you can do little better than one you can mix yourself. With a bit of practice, a personalized shampoo may be formulated to suit you and complement the personality of your pet. Begin by slowly warming ½ cup (113 g) of soft soap with 1½ cups (355 ml) of water. When the soap is melted, stir in 2½ tablespoons (37 ml) of glycerin. As the mixture cools, add 2 tablespoons (30 ml) of denatured alcohol*

* *See Appendix B for cautions pertaining to this ingredient.*

and a few drops of pine oil (or whatever odor appeals to you or your pet). It might be a good idea to make sure your pet's friends are also pleased with your selection. A friend used musk oil in his formulation for the family dog and an amorous skunk tried to move in with her.

*Cattle Ringworm Control*

I remember as a kid that half the old neighborhood gang had ringworm at one time or another during every summer vacation. We weren't fond of wearing any type of footgear, and constant summer rains created a perfect situation for contacting the fungi responsible for causing the reddened ringlike spots to appear on the skin. We were unaware that our association with our pet goats, horses, and contact with cows during milking was also responsible for possible infection.

The fungus responsible for bovine ringworm parasitizes the surface of the skin, which results in the loss of patches of hair. If not treated, this problem may become chronic and an endemic condition may exist. Rubbing areas used by cattle become infested, which results in continual contamination of noninfected animals. Lice frequently are associated with the fungus which, if left untreated, results in unthrifty animals. Men, women, and children can contract the disease from such sources, or by handling infected animals. It's always a good idea to wash the entire body with a good strong soap if contact is made with cattle infested with ringworm. Rubbing areas and contaminated enclosures should be whitewashed with quicklime and all spots on infected animals should be treated. Where an outbreak occurs in feedlots, "dipping" with recommended fungicides with antibiotic treatment may be required.

For spot treatment, a formula that will work well before the fungus becomes epidemic involves tincture of iodine and an oily carrier. Mix 1 cup (237 ml) of tincture of iodine and 1 cup (237 ml) of mineral oil. Stir and apply directly to the skin with a cloth or piece of cotton. If the problem persists or continues to spread, consult your Cooperative Extension agent or veterinarian. (Humans can contract ringworm from infected dogs and cats or other species which inhabit the soil.)

*Cattle Scabies*

Mange, barn itch, scab, itch, tail scab are common names for the condition which arises from attack of cattle and other livestock by mites. Left uncontrolled, scabies can become a serious problem through weight losses and by not being able to sell and ship infected animals from one state to another. Keeping animal quarters clean and painted with whitewash made of lime is a good sanitary practice that will help retard mite populations. However, when livestock become infected, treatment should be made when first detected. If caught early enough, the infested and surrounding areas need treating. If the mites spread, dipping will be required, an inconvenience but no major chore if facilities are available. A formula that is effective on all but one species of mite that causes scabies is made with the following materials.

In 1¼ pounds (568 g) of unslacked lime, add 1 quart (1 l) of water, making a paste. Sift 2.5 pounds (1117 g) of flowers of sulfur into the paste. Bring 3 gallons (12 l) water to a boil and add to the paste. When mixed, boil for 1½ to 2 hours. Cool and strain. Add 7 more gallons (28 l) of water before using. For best results, apply topically or as a dip after heating to 95-105°F.

Follicular or demodectic mange, caused by *Denodex folliculorum bovis,* is not controlled with this treatment. Consult your Cooperative Extension agent or veterinarian when this mange is suspected.

*Dehorning Salve*

Dehorning livestock is a disagreeable chore at best. No matter what process you use to remove the horn, it's always been an unpleasant part of roundup as far as I'm concerned. To prevent infection, or worse, the development of insect pests such as screw worms, a dehorning salve is necessary. After cauterizing, a compound should be used to prevent the recurrence of bleeding, and to eliminate infection by pests. It is a good husbandry practice and can save you money by preventing the development of unthrifty cattle.

Many ranchers have their own effective formulas for a dehorning salve, some of which can become quite laborious to mix. We used to have a first-class hand who always insisted on making his famous brand which came from a hand-me-down recipe from his relatives in Sonora, Mexico. It requires a bottle of tequila as one of the ingredients but I was always suspicious as to how much ever got into the compound. While effective, I was never privileged to obtain the secrets that went into it and can only offer the following as an adequate substitute. I recommend a "copa" of tequila to the ranch hands rather than the dehorning salve.

In 10 pounds (4.5 kg) of pine tar mix 1 pound (454 g) of tannic acid. When well blended, apply liberally to the stump of the horns. When large herds are together, the compound may be rubbed off and should be reapplied as necessary.

### *Dog Foot Conditioner*

The majority of our population is centered in urban communities. Our pets are confined in these same areas where they may be housed within their owners' homes or a small backyard which may or may not have an enclosed kennel with a reduced exercise area. In many cases, it is limited to a walk down the sidewalk on a leash. Dogs, especially, can get out of physical condition quite rapidly and unless we think about it, the working dog may go afield the opening day in poor condition that may result in a gallant try but disappointing performance. Besides poor muscle tone, feet are the principal problem, as inactivity has caused them to become tender. The same problem can develop for the "companion animal" who is taken on an outdoor vacation where the terrain is rough on the tender pads of the feet.

Back in the days when more time was available, I had the opportunity of working my bird dogs in the fields a couple of times a week. Their feet were well calloused, and when hunting season opened in the fall, I never worried that the discomfort of injured feet would reduce their hunting efficiency. Today, however, my time for those enjoyable runs is greatly reduced. So when hunting season rolls around, I use a foot conditioner that toughens the dog's pads to withstand the rigors of rocks, thorns, and sharp grasses.

A satisfactory formula can be made simply at home. It takes but a few minutes to make and less time to apply during repeated applications to the dog's feet.

Tannic acid (used in tanning leather) will toughen the pads on a dog's feet when used alone or in combination with other ingredients. Tannins may be obtained from the bark of hardwood trees. Husks of walnuts are also an excellent source. Place approximately ¼ pound (114 g) of bark or nut husks in 1 quart (1 l) water and boil for 20 to 30 minutes. Use an old pan or suitable can as the tannins will stain the container. If you do not have the opportunity to collect such materials, a cheap grade of black tea can be used with the same results. In ¼ cup (59 g) of tea, add 1 cup (237 ml) of water. Boil 15 minutes to remove the tannins. Allow the water to evaporate until ¼ cup of liquid (59 ml) remains, and strain. Add 1 tablespoon (15 ml) of tincture of benzoin and 1 teaspoon (14 g) of alum. Mix the ingredients together and shake well before applying to the pads of the dog's feet each day for at least 2 weeks prior to its extensive use in the field. You (and the dog) will be satisfied and happy that you took the time to maintain your pet's health.

### *Earwax Remover for Pets*

Those of us who have domestic pets that have given birth to litters of young are aware that Mother Nature provided them with their private bath system. The mother cat, rabbit, or dog spends many of her waking hours thoroughly licking her young ones. I've spent many enjoyable hours watching some of my mother German shorthair pointers going from one end to the other of each puppy, literally lifting them from their warm beds as they put a lot of effort into having clean babies. As they grew older, they became not too much unlike my own children when it came to bath time—they were pretty hard to round up for their share of cleaning. After a period, the mother decides just as we parents, that the training period is over and the social pressures of good sanitation are left to the individual. With effort the process is mastered and our pets can do a reasonable job of keeping soil from their bodies, although we frequently give a helping hand in reducing unwanted body odors.

One area that our pets cannot adequately clean is their ears, and odors and even more serious problems can develop. Because of this, I have gotten into the habit, when I give my pets a bath, of taking a few extra minutes to wash their ears with a practical earwax remover, thereby ensuring that no problems will arise. The formula is simple and inexpensive to make, and to date, I've never had a serious ear problem, even in humid areas of the country where ear cankers can be a hazard to health.

Obtain a bottle of isopropyl alcohol* from the drugstore and to ¼ cup (59 ml) add 10 drops of glycerin that can be obtained from the same location. Store in a clean glass or plastic container. For use, apply to the inside of the ear with a syringe and clean out very carefully with a cotton swab. The usual head shaking will clean out most of the wax. I always have had better cooperation with the pet when the solution was at body temperature. It goes without saying, never place the remover in the ear if it is too cold or hot. These few extra minutes may prevent serious and uncomfortable problems.

### *Manure Fertilizer Activator*

More and more people are trying their hand at gardening, and the reasons are varied. One that is frequently heard is the intent to get away from man-made products such as fertilizers, pesticides, and preservatives. More power to them, if they have the time, space, and energy to enjoy the demands of "organic gardening." In the process, a question that frequently arises pertains to composting or the "breaking down" of organic materials, especially manures from livestock. Usually, there appears to be a sense of urgency in accomplishing this, and where there is, the following activator assists in the decomposition process.

To 1 bushel (8 kg) of livestock manure add 1 package (14 g) of dried yeast in which has been mixed 1 pound (454 g) of brown sugar and 1 gallon (4 l) of warm water. Mix and work in well before spreading a 1-inch layer over the garden soil. Work it well into the soil with a shovel or rototiller, allowing several weeks before planting (if possible).

** See Appendix B for cautions pertaining to this ingredient.*

## NOTES

Name of Formula: ________________________________
Date Made: ____________________
Ingredients and amounts: __________________________
________________________________________________
________________________________________________
________________________________________________

Label: Ingredients and caution warnings
Observations: ___________________________________
________________________________________________
________________________________________________
________________________________________________
________________________________________________

----------

Name of Formula: ________________________________
Date Made: ____________________
Ingredients and amounts: __________________________
________________________________________________
________________________________________________
________________________________________________

Label: Ingredients and caution warnings
Observations: ___________________________________
________________________________________________
________________________________________________
________________________________________________
________________________________________________

## NOTES

Name of Formula: ____________________

Date Made: ____________

Ingredients and amounts: ____________________

____________________

____________________

____________________

Label: Ingredients and caution warnings

Observations: ____________________

____________________

____________________

____________________

____________________

- - - - - - - - - - -

Name of Formula: ____________________

Date Made: ____________

Ingredients and amounts: ____________________

____________________

____________________

____________________

Label: Ingredients and caution warnings

Observations: ____________________

____________________

____________________

____________________

____________________

# 5

# Sports and Camping

## Sports and Camping

## History of Gunpowder

History does not reveal just who was responsible for the discovery of gunpowder; however, we know that it was being used in the early twelfth century by the Chinese, Arabs, and the people of India. The first gun was around 1250, but the western world was without the propellant or the weapon until an English friar, Roger Bacon, published a book in which he described the process of making gunpowder. It was the first explosive to be used both for blasting and to drive shot from guns. In the latter use, it had the disadvantage of producing clouds of smoke when it exploded. Most gunpowder today is used in fireworks, detonators for other explosives, and as an ignitor for rockets.

A German monk, Berthold Schwarz, in the 1300s advanced the development of practical explosives for firearms, and by 1346, cannons had been developed that could hurl large missles at a target some distance away. Until that time, one usually saw his enemy eye to eye, but gunpowder made killing, either for food or defense, somewhat impersonal; and we have been "improving" on the process. Today, there is justified concern that we may have the power to annihilate the human race.

Gunpowder is simple to manufacture and chemically simple to formulate. When we refer to it, we are talking about "black powder," which consists of a mixture of finely divided solids, the percentage of each depending on the intended use of the final product. Powders for guns (and for the muzzle-loaders that have been revived with such popularity) consist of 75-78% potassium nitrate,* 12-15% carbon, and 9-12% sulfur. Blasting powder, for which larger volumes of gas are required, has increased carbon (14-21%) and sulfur (13-18%). When 1 grain of powder is exploded, 250-300 milliliters of gas are formed and 500-700 calories of heat are produced. "Safety" explosives, made from black powder, are still used in mining to prevent igniting mine gases, since the heat generated is less than that produced by high explosives.

The smokeless gunpowder that is used in modern firearms, including shotguns and rifles used for hunting, was not developed until 1884. The advantages of this new powder were obvious, but the higher pressures generated by it (compared with

** See Appendix B for cautions pertaining to this ingredient.*

black powder) necessitated new guns that were capable of withstanding such pressures. The twist-steel and damascus barrels on the scatterguns were designed for black powder, and many a fine gun was put to rest when sporting ammunition went entirely to the smokeless-type powders. Even today, individuals press their luck by using modern shells in these old and handsome guns. I've been on a few hunts where they were used — and totally uncomfortable during the entire time we were out.

Smokeless gunpowder followed the discovery of gun cotton, when in 1846 C. F. Schonbein, a German chemist, developed the process of treated cotton or purified wood cellulose in a mixture of sulfuric acid and nitric acids. Besides gun cotton, nitroglycerin and trinitrotoluene are used in various combinations to produce powders for firearms. The greater pressures produced by nitro-type powders (over black powder) greatly increased the velocity and, consequently, the accuracy of our rifles and shotguns. Another advantage is the lack of the "smoke screen" after each shot, which gives the shooter the advantage of seeing if he hit the target or whether another round is required. There is nothing to indicate a replacement for smokeless powder as the propellant for modern ammunition.

## *Alcohol as a Solid Fuel*

How many times have you found yourself faced with the necessity of building a fire when everything in sight is wet? I've hunted in most of the United States, Mexico, and Canada, and almost every place I've been, sooner or later I'm faced with the problem of using wet wood for a fire. Way back when, I used to come in from the field just dragging tired, and if everything was wet, I'd eat something out of a cold can and call it a meal. But as I grew older, I demanded more, and it appeared to me there should be an easy way to carry a fire-starting compound on even the most demanding backpack trips.

While many may not regularly need to start fires from wet wood, you may wish to have a fuel source for other reasons. If you use chafing dishes or warming plates, this formula should be of interest to you as well as to the outdoor cook, whether in the backyard or in a wilderness area.

Dissolve ½ cup (104 g) of stearic acid in 1 quart (1 l) of denatured alcohol.* Then dissolve 1 tablespoon (14 g) of caustic soda* in another quart (1 l) of alcohol.* Warm both solutions to 140° F. and mix them together. Pour into suitable containers, cool, and cover to prevent evaporation.

WARNING: Caustic soda is a strong irritant and should not come in contact with the skin; flush with clean water if contact is made. SECOND WARNING: You are aware that alcohol will burn, so warm to 140° F. in a *closed* container which has been placed in a pan of warm water, avoiding contact with open flame.

To use, remove the cover, ignite, and place under the object to be heated or ignited. You may save yourself a lot of misery by taking several of these bits along on your next outing.

### *Preparing Bird Skins for Mounting*

My first attempt to mount a bird involved a horned owl. It was so beautiful, I could not bear the thought of its deterioration and decided to save it for posterity by mounting it for the world to see. As I remember, I couldn't have been more than 12 years old and was possessed with the thought that you could do anything if you really wanted to. A wiser philosophy might have prevailed in this case, but no books were available — nor was there a taxidermist anywhere near from whom to inquire for a method of preserving the bird.

My efforts, while gallant, went unrewarded, as the project turned out a disaster. I can't remember what all I used on the skin for tanning, but (at the suggestion of someone) I included hickory-wood ashes mixed with salt. How simple it would have been if I had the formula I use today (as do most taxidermists) in the process of mounting birds.

After the skin has been carefully removed, all tissue and fat must be removed even more carefully. The skin is paper thin in several places and is very easily torn. When cleaned, a small soft-bristle brush can be used to paint the flesh side with a mixture of 4 tablespoons (60 ml) of carbolic acid* and 2 cups (474 ml) of water. Using gloves, dust with borax to prevent insect infection.

** See Appendix B for cautions pertaining to this ingredient.*

## *Canvas Cleaner*

While synthetic fabrics have largely replaced the cotton canvas that was once standard material for all tents, there are a lot of us who still own equipment made from canvas. When I go on a serious hunting trip, I make sure that the old rectangular sheepherder-style tent is among the first equipment I pack. It's strong, wind- and waterproof, and heavy. However, it's only "packed in" when horses are available or when a vehicle can reach the campsite.

Along with being heavy, the canvas tent is prone to getting dirty. Both for aesthetic reasons and for better waterproofing, canvas should be cleaned. Also, cleaning can assist in reducing allergies that may be brought about by accumulated dust in the tent material. I've found a relatively simple and inexpensive way to keep my tent clean. You may wish to try it, and find that it is all you need to eliminate the need for a new tent. With the savings, invest in a new gun or similar item.

Stir 3 tablespoons (42 g) of calcium hypochlorite* in 1 quart (1 l) water until dissolved. Sponge onto the soiled areas, rub in, and rinse off with clear water. *Remember,* calcium hypochlorite is *toxic* and should be treated with caution. Read and follow the label on its container.

## *Darkening Bleached Antlers*

The process of aging and exposure to the elements can cause the antlers from that trophy buck or bull elk to lighten. In time, they can become as white as bone, which distracts from their natural beauty. It's always a good idea to keep mounted heads from direct sunlight, since the hair as well as the antlers begin to fade. I also urge that you find a better place for displaying your heads than over the fireplace. While many pictures depict this as an appropriate place, nothing could be further from the truth for maintenance of mounted specimens. First, there is frequently smoke, dust, and soot which may escape and collect on the hair and antlers; second, a lot of adhesives that go into the mounting process will be dried out and crack; and third, the best hide-

** See Appendix B for cautions pertaining to this ingredient.*

tanning process cannot withstand heat, which leads to its deterioration.

Bleached antlers can be restored to their natural beauty with a little time and effort, and for a few cents.

To 1 pint (473 ml) of water add potassium permangate crystals until a saturated solution is obtained. You will know when this point is reached by observing that no more crystals will dissolve and go into solution. Stir with a wooden stick or spoon as you add a few crystals at a time. The quantity of potassium permangate* will depend upon the mineral quantity of the water. Usually, 2 ounces (56 g) will be sufficient.

Paint this on the antlers, using a rag or a paint brush. The first applications have a violet color, which will turn to a brownish natural color in time. Apply the first coat and in a day or two check the color, adding more if not dark enough to suit your taste.

Always make sure loose dust and grease have been removed before starting the staining operation. This may be done with a rag containing soap and water, with the addition of a little household ammonia. Rinse with clear water and, while wet, add the first coat of the formula.

When the antlers are the desired color, use a clear furniture wax in a light application to keep the atmosphere from the surface. This is much more natural in appearance than the shiny surface that results from using shellac or a similar material. The wax also makes it easier to remove dust with a soft cloth.

With a little patience and resulting skill, you'll be amazed at how beautiful an old bleached pair of antlers can be made to look. However, do not use the same process on the horns of antelope, big horn sheep, or goats.

### *Fireproofing Synthetic Fabrics*

A lot of outdoor equipment in use today is made from synthetic fabrics. These are lighter in weight and can withstand the elements better than products made entirely from natural fibers. They are attractive and quite versatile; however, one problem is retaining the fireproofing that is impregnated into the fabrics

** See Appendix B for cautions pertaining to this ingredient.*

during manufacturing. Constant exposure to the elements may remove the chemicals and necessitate retreatment to ensure the safety of the equipment and its user.

A simple process that is effective and easy begins by adding 1 cup (227 g) of boric acid to 1 gallon (4 l) of water. Soak the fabric in the mixture, wring out, and hang up to dry.

Re-treat fabrics after each laundering. This may be done by adding 1 cup (227 g) of boric acid to the final rinse cycle of the washing machine.

### *Gun-Bluing Compound*

I finally got around to taking "Old Smoker" to the gunsmith for a bluing job. I did so at my son's insistence and because he was embarrassed each time I showed it to people. I was pleased with the finished product but amazed at the price of the work — only $33 less than I paid for the new automatic 12 gauge back in 1942. With two more guns needing the same treatment, I couldn't afford to invest such money and decided to make my own formula and do the job myself. I can't say that the guns I blued look as good as the one done by the gunsmith, but the job is entirely adequate and the money I saved was added toward an elk-hunting trip to Montana. (It's always so easy to rationalize where the financing is coming from for such trips.) You might wish to try the same method for your bluing problems. If you're not satisfied, you don't have much involved and you can then take your gun to a gunsmith for him to blue.

Dissolve ¾ cup (170 g) of ferric chloride, ½ cup (113 g) of antimony chloride,* and ½ cup (113 g) of tannic acid in 1½ cups (355 ml) of water. Before using, make sure the metal is clean of all oils and grease by using a good solvent. Apply 2 or 3 coats of the formula, allowing it to penetrate and dry before each application. Soak a rag in linseed oil and rub down.

WARNING: Antimony chloride is very toxic and should be used with *extreme caution*. Because of this, use the material needed and dispose of the remainder, rather than store it.

** See Appendix B for cautions pertaining to this ingredient.*

*Gun Solvent*

Back in the days of black powder and gun barrels that were not chromed, it was necessary to clean the gun with a good solvent each time it was fired to prevent pitting of the metal. I've seen some beautiful antique shotguns that were immaculate on the outside, but the inside of the tubes looked like termites had nested there. Even with the modern ammunition and metals used in our firearms, it's still a good idea to clean the bore with a solvent after extensive shooting. I thoroughly recommend it for the ardent skeet or trap shooter, who can burn a lot of powder during a day's shooting.

If you're not in the habit of running a rag with solvent through the barrel (or barrels) after each use, at least do so before storing the gun at the end of the season or shooting period. This is more important for older guns than for the newer models, which have chromed barrels. Use a good commercial "nitro" solvent, available from most hardware and sporting goods stores. If you are really into shooting or are a gunsmith and need a large quantity of material, you may wish to make your own solvent. If so, follow this formula and you'll have a satisfactory product that will keep those barrels in first-rate condition.

Mix 2 tablespoons (30 ml) of amyl acetate* and 2 tablespoons (30 ml) of benzene* in 5 tablespoons (60 ml) of motor oil (SAE 30). Use by swabbing the barrels following firing.

WARNING: Amyl acetate is highly flammable and should be used and stored accordingly.

*Insectproofing for Taxidermy Specimens*

In the early days of taxidermy and not until too long ago, arsenic was one of the basic ingredients used in the art of skin preparation. The process was also expected to insectproof the mounted specimen, and it did a good job — for a period of time. But in spite of its effectiveness, unless other precautions were taken, after several years some really fine trophies were lost to the

* *See Appendix B for cautions pertaining to this ingredient.*

attack of various insects. How many taxidermists were also "lost" through association with the highly toxic and poisonous chemical will probably never be known.

The expression "crazy as a hatter" relates to those exposed to arsenic in the preparation of furs that were used in the manufacture of the hats of early periods. Constant exposure brought mental impairedness and apparently became a "trademark" of many in the industry. I know some slow taxidermists who may be suspected of using the chemical, were it not for a more effective and far less toxic compound employed in the art of modern taxidermy.

Before carpet beetles or other insects have the opportunity to attack and ruin your trophies, you can — inexpensively and effectively — clean and insectproof bird and animal mounts by making the following simple and nontoxic formula.

Make a saturated solution with borax in water. (This borax is available in the laundry and bleach sections of your markets.) Keep adding the borax until no more will dissolve, drain the dissolved portion, and apply to skins — as well as the finished mounted specimens — by spraying or carefully wiping with a sponge soaked in the solution.

This process will keep carpet beetles from attacking the skins, and it is nontoxic to those who handle it.

*Leather Preservative (I)*

Of all my outdoor equipment, the most important to me are my leather boots. I have two pair that go along on every outing and, when possible, I alternate them each day. That way they have a chance to dry both inside and out. Now, I've never been one for frills when I reach the outback (though some may think extra shoes and boots are surplus for a trip), but if I'm depending on my feet to get me where I'm going, then I like to give them the kindest treatment I can, for when they hurt from callouses, blisters, and just being hot, the fun is gone. Besides, I've found out over the years that boots will last twice as long with this treatment. To ensure that the leather remains soft, I stick a small can of preservative in the "extra" boots, and when I switch footwear, I take a few minutes to apply it to them. An outstanding formula that you can make for this purpose takes only a few minutes to prepare and a supply will last a long time.

Mix 4 tablespoons (60 ml) of oleic acid and 1 teaspoon (5 ml) of triethanolamine* with 1 cup (237 ml) of neatsfoot oil. Slowly stir in 1 cup (237 ml) of water, then add ¾ cup (178 ml) more of the neatsfoot oil and another ½ cup (118 ml) of water. Stir until thoroughly mixed and place in appropriate containers. I like a plastic bottle, as there is no rusting or breakage. To use, apply with your hand or use a cloth. Warm the mixture for maximum penetration.

### *Leather Preservative (II)*

I'm not sure that anyone has ever indicated that neatsfoot oil is derived from any type of "neat feet," although it is obtained by a rendering process that removes the oil from the shin and feet bones of livestock. It has an affinity for leather, and when diluted with a "carrier" is an excellent preservative for products made from animal hides and pelts. Many of you who are lucky enough to have a fishing creel made of reed and reinforced with leather stripping and hinges appreciate the influence of water on those parts. Keeping the leather soft and pliable is important in retaining your creel in serviceable order.

However, if you're not a fisherman (woman? person?) but still have leather items, such as boots, harness, or saddles, that are constantly exposed to the hazards of the weather, you might wish to mix a formula which is easy to make and as effective as the more expensive types in the local department store.

To 1½ cups (355 ml) of neatsfoot oil add the same quantity of castor oil. Apply liberally with a cloth or sponge; let it penetrate for several minutes; then rub with a clean cloth.

In the event you've been negligent in cleaning your leather items and they've become hardened, don't throw out that favorite pair of hunting boots until you've tried wrapping them in heavy cloth saturated with the leather preservative. Leave for several weeks. You'll be pleasantly surprised at how the leather responds to this treatment. Many times you can be back in the field

* *See Appendix B for cautions pertaining to this ingredient.*

wearing your old, favorite boots. Sure saves the expense of a new pair, and the agony of breaking them in.

*Leather Waterproofing*

For years I went on many enjoyable trips into the Southern swamps in pursuit of game, fish, fowl, or whatever adventure might be available at that particular season of the year. Most of this time I had wet feet, in spite of all efforts to follow the directions of everyone who was good enough to offer a method for waterproofing leather. I remember one that was rather satisfactory, but when the weather turned warm, the boots turned rancid and had to be thrown away. It was quite disappointing, as I had worked hard collecting the possum fat that was reputed to be the secret ingredient in the mixture. I don't remember the other things that went into it, but I remember the smell until this day. Every hound in the neighborhood would go into full bay when I got near. Needless to say, I wasn't too popular with my family or neighbors.

Now, many years later, I've come across a new mixture that works far better and with much less smell than that famous waterproofing recipe that almost lost me a home. You can obtain both ingredients at most hardware stores and, with a few minutes' time, mix a compound that is about as effective for waterproofing leather as you can buy in any commercial preparation.

Mix 1 tablespoon (15 ml) of silicone oil into 1 cup (237 ml) of Stoddards solvent.* Use a soft cloth or small sponge to rub on *liberally*, making sure the area around the stitching is well treated. Repeat as necessary. Store the excess in a suitable container and label for identification.

*Lens Cleaner*

Instruments with fine lenses, such as binoculars, cameras, telescopes, microscopes, and even eyeglasses, should be properly cleaned to prevent foreign particles from scratching them. The abrasive action of the smallest dust particles can have a

** See Appendix B for cautions pertaining to this ingredient.*

devastating effect on precisionground glass surfaces. A camera lens that is scratched usually has to be replaced, which is a costly substitute for careful cleaning. I use a lot of optical equipment in my profession and find that it pays to use a good lens cleaner.

A formula that I find satisfactory consists of ½ cup (113 g) of potassium oleate, ¼ cup (59 ml) of glycerin, and 10 drops (1 ml) of turpentine.* Melt the first two ingredients in the top of a double boiler, turn off the heat, and stir in the turpentine. Cool and pour into a suitable container, label plainly, and date. Apply with a camel-hair brush and remove with a soft, lint-free cloth, face tissue, or lens paper. Fingerprints and the oils associated with them are readily removed by this cleaning process.

### *Magic Writing Surface*

Keeping a gang occupied in camp can become a chore for Mom and Dad or the counselor, especially if bad weather strikes. It takes a lot of imagination to come up with ideas that haven't been tried before. I've been stuck in a mountain tent for a full week, due to rain, and after the second day found myself reading the labels on the food cans backwards to keep myself amused. If you think something like this could happen in the future, you might consider making "magic writing surfaces" for the younger set. Making them is simple, but it takes some time in getting them together. This adds to the anticipation of using the surfaces, which are magic by the fact they can be erased. You might find both making and using the finished product a unique trick to call upon if you've tried everything else before.

You'll need to melt 4 tablespoons (56 g) of beeswax, 9 tablespoons (135 ml) of venice turpentine, 4 tablespoons (56 g) of lard, and 2 tablespoons (30 ml) of mineral oil in the top of a double boiler. When melted and thoroughly mixed, blend in 1 tablespoon (14 g) of carbon black and 3 tablespoons (42 g) of powdered china clay. While hot, paint a uniform coating on a piece of fiberboard or cardboard of a desired size. When cool, place a thin sheet of transparent plastic over the surface. With staples or clear tape, fasten one end, and you're ready to write. (Wax paper can be substituted, but it tears easily.) Use any object for writing that

** See Appendix B for cautions pertaining to this ingredient.*

will not tear the plastic. When you're ready to erase, simply lift the cover from the painted surface and the writing disappears.

### *Tanning Snake Skins*

Having grown up in the Southern swamps, encountering snakes was about as common as finding grits on the table at breakfast. While I usually attempted to give all species a wide berth, there were times the situation demanded more drastic action. The copperheads and rattlesnakes had beautiful skins and I worked hard to preserve them, but I never was totally successful. They either spoiled (with ensuing smell) or became as hard as a board. Once my efforts were rewarded, but carpet beetles invaded my trophy in less than a month and in another few months there wasn't much left.

Many years later, I learned from a good taxidermist friend of mine a simple, fast, and effective method of skinning and tanning snake skins. I still practice the "live and let live" principle when it comes to poisonous snakes; however, the situation arises when it is necessary to kill a rattlesnake—on the ranch, say, to protect pets and livestock, as well as some unsuspecting soul who couldn't cope with it.

If you find it necessary to kill a poisonous snake and wish to save the skin for a hatband (or, if large enough, a belt), you may follow this suggestion for skinning and tanning.

I don't advise trying to skin the head—unless, of course, you're going to have the animal mounted. It's dangerous, as you can puncture yourself with the fangs. Either tape or nail the head at a sufficient height so that the tail doesn't touch the ground. Cut through the skin around the base of the head to the muscle structure, but do *not* cut through the latter — especially on the stomach side, which has a very delicate covering layer of muscles. Then wrap a dry cloth around the cut area. With a gentle downward pull, the skin will peal back as gently as a sock.

It will be necessary to cut around the external openings, as well as through the vertebra at the base of the rattles (if it happens to be the kind that has the built-in "buzzer" system). Next, scrape the tissue and fat with a spoon or dull kitchen knife from the flesh side. Most skins are fairly delicate, so don't apply too much pressure. A very sharp knife should be run down the center of the stomach side to make a straight and even cut. When spread, the sides should be equal.

Next, using tacks or, better, a staple gun, attach the skin to a board of sufficient width to accommodate it. Always keep the flesh side up. Stretch the skin to its full width and length, retaining its natural size. Avoid overstretching, as the skin will tear as it dries.

Dissolve 3 tablespoons (42 g) of borax in ¼ cup (48 ml) of water. Add this to 2 cups (474 ml) of alcohol (isopropyl)* and 2 cups (474 ml) of glycerin. Put in glass jar and label it as to its contents. With a small, soft-bristle paint brush, paint the flesh side with the mixture. Add a light coat each day for a week to 10 days, or until the glycerin no longer penetrates the skin. Whatever excess remains may be removed by wiping with a rag that has been wetted with alcohol.

The scales on the outer side usually remain attached, depending upon the time of year the snake was killed. If they come off easily, remove the scales and apply a few light coats of the formula to bring back the natural color.

### *Tennis/Badminton Net Preservative*

We Americans tend to take our outdoor recreation quite seriously. A sport will become popular for a while, only to be succeeded and replaced by another sport. At present, tennis is very popular, and courts are constructed at an amazing rate to keep up with public demand. Badminton and volleyball are also popular. Each of them has one thing in common: a net over which the ball or shuttlecock must pass. The composition of these nets varies, but many are made of natural fibers. Since they are exposed to the elements night and day for extended periods, they will deteriorate if care is not taken to prevent mildew and dry rot. This can be done quite simply if the following preservative is used for periodic treatments.

Mix ¾ cup (170 g) of copper naphthenate,* ¼ cup (50 ml) fuel oil, and ¾ cup (177 ml) of creosote into 3 cups (711 ml) of naptha.* Saturate the net in the solution and then hang it up to

** See Appendix B for cautions pertaining to this ingredient.*

dry. The treatment should be effective, even under the most severe weather conditions, for 5 to 6 months. WARNING: This is a flammable mixture. Some people are allergic to creosote; avoid skin contact.

*White Leather Dressing*

White leather may be found on luggage, sporting goods, and clothing. Finding a good dressing for leather goods isn't a chore for most colors, but try finding something that is effective and exclusively made for white leather! Those who have encountered this dilemma may be pleased to know you can make a formula yourself that will be highly effective in restoring leather goods which are finished in white.

Slowly heat ⅞ cup (210 ml) of water and add 5 tablespoons (70 g) of titanium dioxide, ½ teaspoon (2 g) of stearic acid, and 1 teaspoon (4 g) of trisodium phosphate,* stirring until completely dissolved. Cool and pour into suitable containers. Apply to leather goods with a cloth or small sponge for cleaning and restoring them to their original condition.

* *See Appendix B for cautions pertaining to this ingredient.*

## NOTES

Name of Formula: ________________________________

Date Made: ____________________

Ingredients and amounts: __________________________

________________________________________________

________________________________________________

________________________________________________

Label: Ingredients and caution warnings

Observations: ___________________________________

________________________________________________

________________________________________________

________________________________________________

________________________________________________

- - - - - - - - - -

Name of Formula: ________________________________

Date Made: ____________________

Ingredients and amounts: __________________________

________________________________________________

________________________________________________

________________________________________________

Label: Ingredients and caution warnings

Observations: ___________________________________

________________________________________________

________________________________________________

________________________________________________

________________________________________________

## NOTES

Name of Formula: ____________________

Date Made: ____________

Ingredients and amounts: ____________________

____________________

____________________

____________________

Label: Ingredients and caution warnings

Observations: ____________________

____________________

____________________

____________________

____________________

- - - - - - - - - -

Name of Formula: ____________________

Date Made: ____________

Ingredients and amounts: ____________________

____________________

____________________

____________________

Label: Ingredients and caution warnings

Observations: ____________________

____________________

____________________

____________________

____________________

## NOTES

Name of Formula: ________________________________

Date Made: ____________________

Ingredients and amounts: ____________________________

______________________________________________

______________________________________________

______________________________________________

Label: Ingredients and caution warnings

Observations: ________________________________

______________________________________________

______________________________________________

______________________________________________

______________________________________________

- - - - - - - - - -

Name of Formula: ________________________________

Date Made: ____________________

Ingredients and amounts: ____________________________

______________________________________________

______________________________________________

______________________________________________

Label: Ingredients and caution warnings

Observations: ________________________________

______________________________________________

______________________________________________

______________________________________________

______________________________________________

APPENDIX A

# Conversion Equivalents

| | | |
|---|---|---|
| 3 teaspoons | equal | 1 tablespoon |
| 2 tablespoons | equal | 1 liquid ounce |
| 4 tablespoons | equal | 1/4 cup |
| 16 tablespoons | equal | 1 cup |
| 2 cups | equal | 1 pint |
| 2 pints | equal | 1 quart |
| 4 quarts | equal | 1 gallon |
| 16 ounces | equal | 1 pound |

## METRIC CONVERSIONS

We are approaching the time when the metric system will phase out our conventional system of weights and measures. But this is a confusing transition to make; so many people are wisely beginning to learn it now. To aid in their effort, the following tables are included, and the proportions listed in each formula are expressed in both systems. Thus, by association, learning the equivalents is far easier. For simplification of measurement, the metrics have been rounded to 1 decimal point.

## CONVERSION FORMULAS

*Gallons into Pounds* — Multiply 8.33 (wt. 1 gallon of water) by the specific gravity (sg) and the result by the number of gallons. (See any chemical dictionary for the sg of a particular chemical.)

*Pounds into Gallons* — Multiply 8.33 by the sg and divide the number of pounds by the result.

*Milliliters into Grams* — Multiply the number of milliliters by the sg.

## Appendix A

*Grams into Milliliters* — Divide the number of grams by the sg.
*Milliliters into Pounds* — Multiply the number of milliliters by the sg, and divide the product by 453.56 (no. of g. per lb.).
*Pounds into Milliliters* — Multiply the number of pounds by 453.56 and divide the product by the sg.
*Milliliters into Ounces* — Multiply the number of milliliters by the sg, and divide the product by 28.35 (no. g. per oz.).
*Ounces into Milliliters* — Multiply the number of ounces by 28.35 and divide the product by the sg.

## CONVERSION FACTORS

*Liquid Measure*

| From | To | Multiply by |
|---|---|---|
| ounces | milliliters | 29.56 |
| pints | liters | 0.47 |
| quarts | liters | 0.95 |
| gallons | liters | 3.78 |
| milliliters | ounces | 0.03 |
| liters | pints | 2.10 |
| liters | quarts | 1.05 |
| liters | gallons | 0.26 |

*Dry Measure*

| From | To | Multiply by |
|---|---|---|
| ounces | grams | 28.35 |
| pounds | kilograms | 0.45 |
| grams | ounces | 0.035 |
| kilograms | pounds | 2.21 |

**FLUID MEASURE**

| Metric | U.S. Regular |
|---|---|
| 1 milliliter | 0.034 ounce |
| 1 liter | 33.81 ounces |
| 1 liter | 2.10 pints |
| 1 liter | 1.05 quarts |
| 1 liter | 0.26 gallons |

**DRY MEASURE**

| Metric | U.S. Regular |
|---|---|
| 1 gram | 0.035 ounce |
| 1 kilogram | 35.27 ounces |
| 1 kilogram | 2.21 pounds |

**DRY MEASURE**

| U.S. Regular | Metric |
|---|---|
| 1/8 teaspoon | 0.54 grams |
| 1/4 " | 1.09 " |
| 1/2 " | 2.19 " |
| 3/4 " | 3.28 " |
| 1 " | 4.38 " |
| 1/8 tablespoon | 1.77 grams |
| 1/4 " | 3.54 " |
| 1/2 " | 7.09 " |
| 3/4 " | 10.63 " |
| 1 " | 14.18 " |
| 1/8 ounce | 3.59 grams |
| 1/4 " | 7.39 " |
| 1/2 " | 14.18 " |
| 3/4 " | 21.34 " |
| 1 " | 28.35 " |

## Appendix A

| U.S. Regular | Metric |
|---|---|
| 1/8 pound | 56.69 grams |
| 1/4 " | 113.39 " |
| 1/2 " | 226.78 " |
| 3/4 " | 340.17 " |
| 1 " | 453.56 " |
| | |
| 1/8 cup | 28.34 grams |
| 1/4 " | 56.69 " |
| 1/2 " | 113.39 " |
| 3/4 " | 170.08 " |
| 1 " | 226.78 " |

## LIQUID MEASURE

| U.S. Regular | Metric |
|---|---|
| 1/8 teaspoon | 0.61 milliliters |
| 1/4 " | 1.23 " |
| 1/2 " | 2.47 " |
| 3/4 " | 3.70 " |
| 1 " | 4.94 " |
| | |
| 1/8 tablespoon | 1.84 milliliters |
| 1/4 " | 3.69 " |
| 1/2 " | 7.39 " |
| 3/4 " | 11.08 " |
| 1 " | 14.78 " |
| | |
| 1/8 ounce | 3.69 milliliters |
| 1/4 " | 7.39 " |
| 1/2 " | 14.78 " |
| 3/4 " | 22.17 " |
| 1 " | 29.57 " |

| U.S. Regular | Metric |
| --- | --- |
| 1/8 cup | 29.57 milliliters |
| 1/4 " | 59.14 " |
| 1/2 " | 118.28 " |
| 3/4 " | 177.42 " |
| 1 " | 236.56 " |
| | |
| 1 pint | 473.00 milliliters |
| 1 quart | 946.00 " |
| 1/2 gallon | 1.89 liters |
| 3/4 " | 2.83 " |
| 1 " | 3.78 " |

APPENDIX B

# Definitions of Chemicals Used in *The Formula Book 3*

ACACIA (Gum Arabic): White powder or flakes, soluble in water. Dried from the plant Acacia Senegal.

ACETIC ACID (Vinegar Acid): Clear colorless liquid, miscible with water, alcohol, glycerin, and other. Made by oxidation of petroleum gases.

ACETONE: Miscible with water, alcohol, ether, chloroform, and most oils. Made by oxidation of cumene. CAUTION: Extremely volatile and flammable.

ACTIVATED CHARCOAL: Black powder. Obtained by the destructive distillation of carbonaceous materials such as wood or nut shells. It is activated by heating to approximately 900° C. with steam or carbon dioxide which produces a honeycomb internal structure, making it highly adsorptive.

ALMOND OIL: White to yellowish oil, distilled from ground kernels of bitter almonds imported from Spain, Portugal, or France. CAUTION: Vapors are toxic.

ALUMINUM CHLORIDE, ANHYDROUS: White or yellowish crystals. Derived by reaction of purified gaseous chlorine with molten aluminum, by reaction of bauxite with coke and chlorine at about 1600° F. CAUTION: Highly toxic by ingestion and inhalation, strong irritant to tissue.

ALUMINUM POTASSIUM SULFATE (Alum): White crystals or

powder, soluble in water. Derived from alunite leucite and other minerals. Acts as an astringent.

ALUMINUM POWDER: Gray to silver powder, milled from aluminum or its alloys. Particles are dispersed in a vehicle such as paint.

ALUMINUM STEARATE: White powder, soluble in petroleum and turpentine oil. Made by reacting aluminum salts with stearic acid.

ALUMINUM SULFATE (Alum): White crystals, soluble in water. Made by treating kaolin with sulfuric acid.

AMMONIA, HOUSEHOLD: A dilute solution of ammonium hydroxide.

AMMONIUM SELENATE: Colorless crystals, soluble in water. CAUTION: May be mildly toxic.

AMMONIUM CARBONATE (Hartshorn): White powder, soluble in cold water. A mixture of ammonium acid carbonate and ammonium carbamate. Derived from the heating of ammonium salts with calcium carbonate. CAUTION: When heated, irritating fumes may result.

AMMONIUM CHLORIDE (Sal Ammoniac): White crystals, soluble in water and glycerol. Derived from the reaction of ammonium sulfate and sodium chloride solutions.

AMMONIUM CITRATE: White granules, soluble in water.

AMMONIUM HYDROXIDE (Aqua Ammonium): Water solution of ammonia gas. CAUTION: Toxic by ingestion. Liquid and vapors may be irritating to eyes and skin.

AMMONIUM OLEATE (Ammonia Soap): Brown, jellylike mass, soluble in water and alcohol. Acts as an emulsifying agent.

AMMONIUM NITRATE (Saltpeter): Colorless crystals, soluble

in water, alcohol, and alkalies. Made by the action of ammonia vapor on nitric acid. CAUTION: Do not store in high temperatures.

AMMONIUM PHOSPHATE: White crystals moderately soluble in water. Derived from the interaction of phosphoric acid and ammonia.

AMMONIUM STEARATE: Tan, waxlike solid, dispersible in hot water, soluble in hot toluene.

AMMONIUM SULFATE: Gray to white crystals, soluble in water. Made by neutralizing synthetic ammonia with sulfuric acid.

AMYL ACETATE (Banana Oil): Volatile bananalike smell. CAUTION: Toxic.

ANHYDROUS LANOLIN (Wool Fat): Brown jelly, miscible with water. Soluble in benzene, ether, acetone, and slightly soluble in cold alcohol.

ANITERACENE OIL: A medium viscosity oil obtained as a coal tar fraction.

ANTHRACENE OIL: A coal tar fraction. CAUTION: Hazardous, toxic, and irritant.

ANTIMONY CHLORIDE: White powder soluble in hydrochloric acid and alkali tartrate solutions. CAUTION: Highly toxic.

ANTIMONY POTASSIUM TARTRATE (Tartar Emeric): White powder, soluble in water. Derived by heating antimony trioxide with a solution of potassium bitartrate and then crystalized. CAUTION: Toxic if taken internally.

ASBESTOS POWDER: Gray fibrous powder. Mined as a natural mineral. CAUTION: Do not inhale dust.

ASPHALT (Residual Oil, Petroleum Asphalt, Trinidad Pitch,

Mineral Pitch): Solid to semisolid lumps, turns to viscous liquid at 200° F.

BALL CLAY: Tan-colored, powdered, highly plastic with strong bonding power. Mined in various sections of the United States.

BARIUM SULFIDE (Black Ash): Yellowish green to gray powder, soluble in water. Made by roasting barium sulfate and coal together, adding water, and evaporating. CAUTION: Toxic if taken internally.

BAY RUM: Amber liquid made from a mixture of bay oil, orange peel oil, oil of pimenta, and alcohol.

BEEF TALLOW: Solid fatty material found in beef.

BEESWAX: See White Beeswax.

BENTONITE (Sodium Bentonite): Light powder, insoluble in water, expands to many times its size in water. Mined in Wyoming, Mississippi, Texas, Canada, Italy, and Russia.

BENZENE: Colorless liquid, made by the catalytic reforming of petroleum, and also by the fractional distillation of coal tar. CAUTION: Flammable.

BENZENE HEXACHLORIDE (BHC, Lindane): Insecticide. CAUTION: Highly toxic.

BORAX (Sodium Borate): White powder, soluble in water. Mined in the western United States.

BORIC ACID (Boracic Acid): Colorless, odorless, white powder, soluble in water, alcohol, and glycerine. Made by the addition of hydrochloric or sulfuric acid to a borax solution, and then crystallizing.

BENZOIC ACID (Carboxybenzene): White scales or needlelike crystals, soluble in alcohol, ether, chloroform, benzene, and turpentine. Made by the oxidation of toluene.

BERGAMOT OIL: Honey-colored oil, soluble in alcohol. Derived from the fruits of *Citrus Bergamia Risso et Painteau.*

BURGUNDY PITCH: A tacky liquid, soluble in alcohol and acetone. Extracted from Norway spruce.

BUTYL ALCOHOL (Butanol): Colorless liquid, soluble in water, miscible with alcohol and ether. Made by the hydrogenation of butylraldehyde, obtained in the Oxo process.

CADMIUM ACETATE: Colorless crystals, soluble in water. CAUTION: Toxic.

CALCINED MAGNESIA (Magnesite): Derived by firing magnesite up to 1450° C., at which time it converts to an adsorptive medium with high internal porosity.

CALCIUM CARBONATE (Chalk): White powder slightly soluble in water, highly soluble in acids. Derived principally from limestone.

CALCIUM CHLORIDE: White flakes that decompose in water. Absorptive agent.

CALCIUM HYPOCHLORITE (Chlorinated Lime): White crystalline solid, soluble in water. Derived from the chlorination of a lime/caustic slurry. CAUTION:Toxic.

CALCIUM LACTATE: White, tasteless powder. Soluble in water.

CALCIUM SULFATE: White, odorless crystals or powder. only slightly soluble in water. Occurs in nature as a hydrated form of gympsum, and also as an anhydrate.

CAMPHOR (Gum Camphor, Camphanone): Colorless or white crystals, soluble in alcohol. Derivation: steam distillation of camphor tree wood. CAUTION: Vapors flammable.

CAMPHOR OIL: Pale yellow oily liquid. Made by distilling the flowers of the Canaga odorata that grows in Java.

CARBON DISULFIDE: Clear liquid, soluble in alcohol, benzene, and ether. Highly flammable.

CARBORUNDUM: Trade name for a full line of abrasives made by the Carborundum Company.

CARNAUBA WAX (Brazil Wax): Yellow to brown hard lumps, melting point 84°-86° C. Collected from the leaves of the Brazilian wax palm, *Copernica cerifera.*

CARBOLIC ACID (Phenol): Soluble in water and alcohol. Made by the oxidation of cumene. CAUTION: Toxic by ingestion, inhalation, and skin absorption.

CARBON BLACK (Stove Black, Furnace Black): Black insoluble amorphous powder. Made by the incomplete combustion of natural gas or petroleum.

CASTILE SOAP: Olive oil is used for Castile soap; transparent soaps are made from decolorized fats.

CASTOR OIL (Ricinus Oil): Pale yellow oil, soluble in alcohol. Derived from pressing the seeds of the castor bean, *Ricinus communis.*

CAUSTIC POTASH (Potassium Hydroxide): White flakes, soluble in alcohol, water, or glycerin. Made by electrolysis of a potassium chloride solution. CAUTION: Heats on contact with water, can cause severe burns to skin. Handle with care. Store in airtight container.

CAUSTIC SODA (Sodium Hydroxide): White chips, soluble in water or alcohol. Made by electrolysis of a sodium chloride solution. CAUTION: Heats on contact with water, can cause severe burns to skin. Handle with care. Store in airtight container.

CEDAR OIL: An aromatic essential oil extracted from cedar bark and wood.

CEDAR OIL EMULSION: Cedar oil emulsified with water, using any suitable emulsifier such as liquid detergent.

CERESIN WAX (Ozocerite, Mineral Wax): White or yellow solid, soluble in alcohol. Melting point 68°-72° C. Made by purifying ozocerite with sulfuric acid and then filtering through charcoal.

CETYL ALCOHOL: White crystals, soluble in alcohol. Melting point 49° C. Made by saponifying spermaceti with caustic alkali.

CETYLTRIMETHYLAMMONIUM BROMIDE (Ammonium Salt): White powder, soluble in water or alcohol. Has surface-active and germicidal properties.

CHLORINATED LIME (Bleaching Powder): White granules that decompose in water. Made by reacting chlorine with slaked lime. CAUTION: Forms chlorine when mixed with water.

CHLOROPHYLL: Green material found in plants and algae. Can be had in aqueous, alcoholic, or oil solutions. Made by extraction from the plant source.

CHROMIC OXIDE (Chromium Oxide, Chromia, Chromium Sesquioxide, Green Cinnabar): Bright-green crystalline powder, insoluble in water, acids, and alkalies. Derived by heating chromium hydroxide; dry ammonium dichromate, and sodium dichromate with sulfur and washing out the sodium sulfate.

CINNAMON OIL (Cassia Oil): Light yellow aromatic oil, soluble in alcohol. Distilled from the leaves and twigs of the plant *Cinnamonum cassia*.

CITRIC ACID: White crystals, soluble in water or alcohol. Derived by mold fermentation from lemon, lime, pineapple juice, and molasses.

CITRONELLA OIL: Light yellow essential oil, soluble in alcohol. Derived by the steam distillation of the grass *Cymbopogon Nardus*. CAUTION: Mildly toxic if taken internally.

CLAY (Hydrated Aluminum Silicate): Tan powder, ranging in particle size from 150 to less than 1 micron. Absorbs water to form a plastic mass. Derived from nature by natural weathering, crushing, and screening of rock.

CLOVES: The dried flowers of *Eugenia aromatica*. Flowers are frequently ground to a powder, or distilled to produce oil of clove.

COAL TAR: Black viscous liquid (or semisolid), napthalenelike odor; sharp burning taste; obtained by destructive distillation of coal. Soluble in ether, benzene, carbon disulfide, chloroform; partially soluble in alcohol, acetone, methanol, benzene; slightly soluble in water. CAUTION: Highly toxic by inhalation.

COBALT CHLORIDE (Cobaltus Chloride): Blue or red crystals. Soluble in water, alcohol, or acetone. Made by the action of hydrochloric acid on cobalt.

COCONUT OIL: White, semisolid, lardlike fat. Soluble in alcohol. Melting point 83° F. Made by press extraction of coconut meat followed by alkali refining.

COCOA BUTTER (Theobroma Oil): Yellow-white solid. Melting point 30°-35° C. Soluble in ether or chloroform. Made by the expression of cocoa beans and solvent extraction.

COPPER NAPHTHENATE: A green-blue solid, soluble in gasoline, benzene, and mineral oil distillates. Made by combining cupric to a solution of sodium napthenate. CAUTION: Mildly toxic by ingestion.

COPPER SULFATE (Blue Vitriol, Bluestone): Blue crystals, lumps, or powder. Soluble in water or methanol. Made by the action of dilute sulfuric acid on copper or its oxides. CAUTION: Highly toxic.

CORN OIL (Maize Oil): Pale yellow liquid, partially soluble in alcohol. The germ is removed from the kernel and cold pressed.

CORN SYRUP (Glucose): Viscous liquid consisting of a mixture of dextrose, maltose, and dextrius with about 20% water. Soluble in water and glycerine. Made by the hydrolysis of starch and the action of hydrochloric acid.

COTTONSEED OIL: Pale yellow to clear oil, soluble in ether, benzene, chloroform, and carbon disulfide. Made by solvent extraction or hot pressing of cotton seeds.

CRESOL (Methyl Phenol): Colorless or yellowish to pinkish liquid. Soluble in alcohol, glycol, and dilute alkalies. CAUTION: Toxic and irritant.

CREOSOTE (Wood Tar, Beechwood): Colorless oil liquid, miscible with alcohol or ether. A mixture of phenols obtained by the destructive distillation of wood tar.

CRESYLIC ACID: A commercial mixture of phenolic materials, made from petroleum or coal tar. CAUTION: Toxic, absorbed through skin.

DENATURED ALCOHOL: Ethyl alcohol that has been contaminated with a minute amount of another material to make it unfit for human consumption as a beverage. Clear white liquid. CAUTION: May be toxic if taken internally. Flammable.

DEODORIZED KEROSENE: Kerosene that has been treated chemically to mask its odor. CAUTION: Toxic if taken internally. Flammable.

DERRIS POWDER: See Pyrethrum.

DERRIS ROOT: The root of the shrub *Malacceusis*. Chief active ingredient is rotenone.

DIBASIC AMMONIUM PHOSPHATE (Ammonium Phosphate, Secondary; Diammonium Hydrogen Phosphate; Diammonium Phosphate; DAP): White crystals or powder, milky alkaline in reaction, soluble in water.

DIATOMACEOUS EARTH (Keiselguhr, Diatomite): A bulky light material containing 88% silica. The balance is made up of the skeletons of small prehistoric plants related to algae. Can be had in either brick or powder form.

DIGLYCOL LAURATE (Diethylene Glycol Monolaurate): Straw-colored, oily liquid, nontoxic. Dispersible in water. Derived from thelauric acid ester of diethylene glycol.

DIGLYCOL OLEATE (Diethylene Glycol Monooleate): Light red, oily liquid; fatty odor. Soluble in ethanol, naptha, ethyl acetate, methanol; partly soluble in cottonseed oil; insoluble in water. Combustible; nontoxic.

DIGLYCOL STEARATE: White, waxlike solid. Disperses in hot water, soluble in hot alcohol. Made by using stearic acid and the ester of diethylene glycol.

DIMETHYLMORPHOLINE: Liquid. Flash point 112° F. CAUTION: Flammable.

ESSENTIAL OILS: Volatile oils derived from the leaves, stems, flowers, and twigs of plants, and from the rinds of fruits. Methods of extraction are by steam distillation, pressing fruit rinds, solvent extraction, and maceration of flowers and leaves. Generally soluble in alcohol and chloroform.

ETHYL ACETATE (Acetic Ether; Acetic Ester, Vinegar Naphtha): Colorless, fragrant liquid. Soluble in chloroform, alcohol, and ether; slightly soluble in water. CAUTION: Moderately toxic by inhalation and skin absorption. Irritating to eyes and skin. Highly combustible.

ETHYL ALCOHOL (VODKA): See Introduction.

ETHYLENE CHLORIDE (Ethylene Dichloride): Colorless, oily liquid, miscible with most organic solvents. Made by the action of chlorine or ethylene. CAUTION: Toxic by ingestion, inhalation, and skin absorption. Irritant to eyes and skin. Handle with care.

ETHYLENE GLYCOL (Glycol): Clear, colorless, syrupy liquid, soluble in water and alcohol. Made from formaldehyde, water, and carbon monoxide with hydrogenation of the resulting glycolic acid.

ETHYLENE GLYCOL MONOETHYLETHER (Cellosolve): Colorless liquid, miscible with water and hydrocarbons. Flash point 120° F. CAUTION: Flammable.

FERRIC CHLORIDE (Iron Chloride): A black-brown solid in water, alcohol, and glycerol. Made by the action of chlorine on ferrous sulfate.

FERRIC OXIDE: Iron mineral appearing natural.

FISH OIL: A drying oil obtained from salt-water fish. Used in soap making.

FLOWERS OF SULFUR: Yellow crystals or powder, partially soluble in alcohol. Melting point 112° C. Mined in various sections of the United States.

FORMALDEHYDE: An aqueous solution. Made by oxidation of synthetic methanol. CAUTION: Highly toxic by ingestion, inhalation, or skin contact.

FUEL OIL (Furnace Oil): Number 1 or 2 grade. Oil used in home-heating furnaces.

FULLER'S EARTH: A porous colloidal aluminum silicate of 1 micron or less, having high adsorptive power. Mined in Florida, England, and Canada.

FURFURALDEHYDE (Bran Oil): Colorless liquid, soluble in water. Derived from grain hulls. CAUTION: Highly toxic, can be absorbed through skin.

GELATIN: White to yellow powder, soluble in hot water. Made by boiling animal by-products with water. Will absorb up to 10 times its weight of water.

GLUCOSE: Starch or corn syrup. Viscous, syrupy liquid. A mixture of dextrose, maltose, and dextrin. Colorless to yellowish, soluble in water and alcohol.

GLYCERIN (Glycerol): A clear, colorless, syrupy liquid, soluble in water and alcohol. Made by the hydrogenation of carbohydrates with a nickel catalyst.

GUAIACOL: Yellowish, oily liquid or crystals. Soluble in alcohol, moderately soluble in water. Made by extracting beechwood creosote with alcoholic potash. CAUTION: Moderately toxic.

HEXACHLORODIPHENYL OXIDE: Light yellow liquid, soluble in methanol ether.

HEXAMETHYLENETETRAMINE: White crystalline powder, soluble in water. Made by the action of ammonia on formaldehyde. CAUTION: Moderately toxic, flammable.

HYDRATED LIME (Calcium Hydroxide): White powder, soluble in glycerin. Made by the action of water on calcium oxide. CAUTION: Skin irritant.

HYDROCHLORIC ACID (Muriatic Acid): Hydrogen chloride in water solution. Derived as a by-product of the chlorination of benzene. CAUTION: Highly toxic by ingestion and inhalation, can be absorbed by skin. Strong irritant to eyes and skin.

HYDROFLUORIC ACID (Hydrogen Fluoride in Aqueous Solution): Colorless, fuming, mobile liquid. Will attack glass and any silica-containing material. CAUTION: Highly corrosive to skin and mucous membranes; highly toxic by ingestion and inhalation. Use with extreme caution. Wear rubber gloves and use in well-ventilated area.

HYDROGEN PEROXIDE: Colorless dilute aqua solution. CAUTION: Highly toxic in concentrated form. Relatively low toxicity in dilute aqua solution. Sold as 3% solution over drugstore counter as a general antiseptic.

INDIA INK: Carbon black with shellac, borax, or soap in water.

IRON CARBONYL (Iron Pentacarbonyl): Yellow liquid, soluble in organic solvents. Made by treating iron dust with carbon monoxide in the presence of the catalyst ammonia.

IRON OXIDE (Jeweler's Rouge): Reddish-brown fine powder, soluble in acids. Made by the interaction of a solution of ferrous sulfate and sodium carbonate.

IRON SULFATE (Ferric Sulfate): Yellow crystals, soluble in water. Made by the addition of sulfuric acid to ferric hydroxide.

ISOPROPYL ALCOHOL (Isopropanol): White, sweet-smelling liquid. Soluble in water, ether, or alcohol. Made by treating propylene with sulfuric acid and then hydrolyzing. CAUTION: Mildly toxic by inhalation and ingestion. Flammable.

JAPAN WAX (Sumac Wax): Pale yellow solid, soluble in benzene and naphtha. Melting point 53° C.

JEWELER'S ROUGE: See Iron Oxide.

JUNIPER OIL (Juniper Tar, Juniper Tar Oil): Thick, clear dark-brown liquid; tarry odor; burning, bitter taste. Soluble in ether, glacial acetic acid, chloroform; partly soluble in alcohol; very slightly soluble in water. Derived by distillation of the wood of *Juniperus oxycedrus*.

KEROSENE: Oily liquid distilled from petroleum. CAUTION: Toxic if taken internally. Flammable.

LACTIC ACID (Milk Acid): Colorless viscous liquid, miscible with water, alcohol, and glycerin. Made by hydrolysis of lactonitrile.

LAMPBLACK: A black or gray pigment made by burning low-grade heavy oils or similar carbonaceous materials with insufficient air, and in a closed system such that soot can be collected in settling chambers. Strongly hydrophobic.

LANOLIN (Wool Fat): Yellow to light-gray semisolid, soluble in ether or chloroform. Extracted from raw wool and refined.

LARD: Purified internal fat of the hog. Soft white unctuous mass, faint odor, bland taste. Soluble in ether, chloroform, light petroleum hydrocarbons, carbon disulfide; insoluble in water.

LATEX (Liquid Rubber): A white, free-flowing liquid obtained from certain species of trees and shrubs. Usually emulsified with water.

LAURYL PYRIDINIUM CHLORIDE: Mottled tan semisolid, soluble in water. CAUTION: May be mildly irritating to skin.

LAVENDER OIL: Essential oil; colorless, yellowish or greenish-yellow; sweet odor; slightly bitter taste. Steam distilled from fresh flowering tops of *Lavandula officinalis.*

LEAD: Heavy soft-gray metal solid, soluble in dilute nitric acid. Made by the roasting of lead sulfide, lead sulfate, and carbonate. CAUTION: Poison.

LECITHIN: Light brown to brown semiliquid, partly soluble in water, soluble in chloroform and benzene. Made from soybean oil, corn oil, egg yolk, and vegetable seeds.

LEMON OIL: Yellow liquid; soluble in alcohol, vegetable oils, and mineral oil. Expressed from the peel of lemons.

LINSEED OIL: Amber to brown oil, soluble in alcohol. Made by refining raw linseed oil. Warning: Dries when exposed to air. Keep in airtight container.

LUBRICATING OIL: Amber to red liquid of varying viscosity, refined from crude petroleum oil.

MAGNESIUM CARBONATE: White, bulky, light powder; soluble in acids, insoluble in water and alcohol. Made by a union of magnesium sulfate and sodium carbonate solutions.

MAGNESIUM CHLORIDE: White, deliquescent crystals, soluble in water.

MAGNESIUM FLUOSILICATE (Magnesium Silicofluoride): White efflorescent crystalline powder, soluble in water. Derived by treating magnesium hydroxide or carbonate with hydrofluosilicic acid. CAUTION: Highly toxic, strong irritant. Handle with rubber gloves in well-ventilated area only.

MAGNESIUM STEARATE: Light, white powder, insoluble in water or alcohol.

MAGNESIUM SULFATE (Epsom Salts): Colorless crystals, soluble in water and glycerol. Made by the action of sulfuric acid on magnesium oxide.

MANGANESE CHLORIDE: Rose-colored crystals, soluble in water. Slightly toxic.

MENTHOL (Peppermint Camphor): White crystals with strong mint odor; soluble in alcohol, petroleum solvents, and glacial acetic acid. Crystals are formed as a result of freezing mint oil.

MERCURIC CHLORIDE: White crystals or powder; soluble in water, alcohol, or ether. CAUTION: Highly toxic by ingestion, inhalation, and skin absorption.

METHYLCELLULOSE (Cellulose Methyl Etiehr; "Methocel"): Grayish white, fibrous powder; aqueous suspensions neutral to litmus. Swells in water to a colloidal solution. Insoluble in alcohol, ether, chloroform, and in water warmer than 123° F. Soluble in glacial acetic acid, unaffected by oils and greases. Derived from cellulose by conversion to alkali cellulose and then, reacting this, with methyl chloride, dimethyl sulfate, or methyl alcohol and dehydrating agents.

METHYL SALICYLATE (Wintergreen Oil): Colorless or yellow or reddish liquid, soluble in alcohol and glacial acetic acid. Made by heating methanol and salicylic acid in the presence of sulfuric acid.

MINERAL OIL, WHITE (Liquid Petrolatum): Colorless transparent oil, distilled from petroleum.

MINERAL SPIRITS (Petroleum Naphtha): Clear liquid, from the petroleum distillation process. CAUTION: Flammable.

MONOCALCIUM PHOSPHATE: See Superphosphate.

MONTAN WAX (Lignite Wax): White, hard, earth wax, soluble in benzene. Melting point 90° C. Made by extraction of lignite from coal.

NAPHTHA (V.M.P.): White, highly volatile liquid, made in the petroleum distillation process. CAUTION: Highly flammable.

NAPHTHALENE (Tar Camphor): White crystalline flakes, soluble in benzene, absolute alcohol, and ether. Made by boiling coal tar oil and then crystallizing.

NEATSFOOT OIL: A pale yellow oil, soluble in alcohol and kerosene. Made by boiling in water the shinbones and feet, without hoofs, of cattle. The oil and fat are then separated.

NEROLI OIL (Orange Flower Oil): Amber-color oil, soluble in equal parts of alcohol. Made by the distillation of citrus flowers.

NICOTINE SULFATE: White crystals, soluble in water or alcohol. Made by the action of sulfuric acid on the alkaloid. CAUTION: Toxic.

OIL SOAP: An emulsion used by machine shops for lubricating metal cutting tools.

OLEIC ACID (Red Oil): Yellow to red oily liquid, soluble in alcohol and organic solvents. Derived from animal tallow or vegetable oils.

OLIVE OIL: Pale yellow to greenish liquid, nondrying. Only slightly soluble in alcohol. Soluble in ether, chloroform, or carbon disulfide. Oil is cold-pressed from the olive fruit and then refined.

OLEORESIN CAPSICUM: Oleoresin is a semisolid mixture of the resin and the essential oil of the plant from which it is derived, *Capsicum* (cayenne pepper, African pepper, red pepper). Dried fruit of *Capsicum frutescens, Capsicum annuum,* or the Louisiana sport pepper.

ORANGE OIL (Citrus Seed Oil): Orange-color oil, expressed from orange seeds. Bitter taste is removed by refining.

ORTHODICHLOROBENZENE: Colorless, heavy liquid, miscible with most organic solvents. Made by chlorinating monochlorobenzene. CAUTION: Moderately toxic by ingestion, but highly irritating to skin and eyes.

OXALIC ACID: Transparent, colorless crystals, formed in nature by the oxidation of proteins in plants such as wood sorrel, rhubarb, and spinach. CAUTION: Toxic.

OXYQUINOLINE SULFATE: Pale yellow powder, soluble in water. CAUTION: Moderately toxic in concentrated form.

PALM OIL (Palm Butter): Yellow-brown soft, solid material, soluble in alcohol. Derived from nuts and fruit of the palm tree native to West Africa.

PARACHLOROMETACRESOL: Colorless to yellow liquid, soluble in alcohol. Derived from coal tar.

PARADICHLOROBENZENE: White volatile crystals, soluble in alcohol, benzene, and ether. Made by chlorination of monochlorobenzene. Moderately toxic by ingestion. Irritant to eyes. Also known as moth crystals.

PANCREATIN: Cream-color powder, soluble in water. Extracted from the pancreas of cattle and hogs.

PARAFFIN OIL: An oil pressed from paraffin distillate. For characteristics, see Paraffin Wax.

PARAFFIN WAX: White, waxy blocks, soluble in benzene,

warm alcohol, turpentine, and olive oil. Made by distilling crude petroleum oil.

PEANUT OIL (Groundnut Oil): Yellow oil, soluble in petroleum, ether, carbon disulfide, and chloroform. Can be saponified by alkali hydroxides to form a soap.

PENTACHLOROPHENOL: White powder or crystals, soluble in alcohol. Made by chlorinating phenol. CAUTION: Highly toxic by ingestion, inhalation, and absorption through the skin.

PETROLEUM DISTILLATE: Colorless, volatile liquid, miscible with most organic solvents and oils. Made by distillation from petroleum.

PEPPERMINT OIL: Clear, oily liquid, soluble in alcohol. Made by distilling the leaves of the peppermint plant.

PETROLATUM (Mineral Wax, Petroleum Jelly, Mineral Jelly): Colorless to amber translucent oil mass, soluble in benzene, ether, chloroform, and oil. Melting point 60° C. Made by the distillation of still residues from steam distillation of paraffin-based petroleum.

PHENOL: See Carbolic Acid.

PINE OIL: Colorless to amber oily liquid. Miscible with alcohol. Made by steam distillation of pine wood.

PINE TAR: Resinous product derived from the sap of pine trees usually obtained as a by-product of turpentine distillation.

PORTLAND CEMENT: White to gray powder composed of lime, alumina, silica, and iron oxide.

POTASH (Potassium Carbonate, Pearl Ash): White deliquescent translucent powder, soluble in water. CAUTION: Toxic if taken internally.

POTASSIUM CARBONATE: See Potash.

POTASSIUM HYDROXIDE: See Caustic Potash.

POTASSIUM NITRATE (Niter, Saltpeter): Transparent or white crystals or powder, soluble in water. CAUTION: Dangerous fire and explosion risk when subjected to shock or heating. Oxidizing agent. Handle carefully.

POTASSIUM OLEATE: Gray to tan paste, soluble in water and alcohol.

POTASSIUM PERMANGANATE: Permanganate and potash, purple solid soluble formed by oxidation of ossified potassium manganate solution. CAUTION: Caustic, do not allow contact with skin.

POTASSIUM PERSULFATE: White crystals, soluble in water. Made by electrolysis of a saturated solution of potassium sulfate. CAUTION: Moderately toxic.

POTASSIUM SULFATE (Potash Alum): White crystals, soluble in water. Derived from the mineral alunite.

POWDERED SKIM MILK: Powdered milk is prepared by dehydrating whole milk.

POWDERED SULFUR: Nonmetalic mined mineral.

PROPYLENE GLYCOL: Colorless viscous liquid, odorless and tasteless. Miscible with water and alcohol. Made by hydration of propylene oxide.

PROPYLENE GLYCOL MONSTEARATE: A mixture of propylene glycol mono- and diesters of stearic acid. Soluble in alcohol, chloroform, and other chlorinated hydrocarbons. White flakes or beads, bland odor and taste. CAUTION: Combustible.

PUMICE POWDER: A gently abrasive fine powder milled from porous rock found in nature.

PYRETHRUM EXTRACT: A powder obtained from milling

ground pyrethrum flowers. Usually mixed with kerosene or other solvents. CAUTION: Moderately toxic if taken internally.

QUASSIA (Bitterwood, Bitter Ash): White to bright yellow chips. Takes on a tannish color when ground. Obtained from the bark of *Picrasma* or *Quassia* trees.

RAPE SEED OIL (Colza Oil, Rape Oil): Pale yellow viscous liquid, soluble in ether, chloroform, and carbon disulfide. Made by expression or solvent extraction of rape seeds.

RAW LINSEED OIL: Untreated oil from the flaxseed presses which is filtered through duck and flannel cloths in a plate and frame press. Yellow-brown or amber in color. Develops heat spontaneously.

REFINED TALLOW: An animal fat extracted from the solid fat or "suet" of cattle, sheep, or horses by dry or wet rendering.

ROSEMARY OIL: Clear to slightly yellowish oil, soluble in alcohol, ether, and glacial acetic acid. Made by steam-distilling the flowers of *Rosmarinus officinalis.*

ROSIN: Translucent amber chips, soluble in alcohol, ether, glacial acetic acid, and oil. Derived by steam distillation of the sap of pine trees.

ROTTEN STONE (Tripoli): White abrasive powder. Crushed and milled from rock.

SAL SODA (Washing Soda, Sodium Carbonate Decahydrate): White crystals soluble in water, insoluble in alcohol. Pure form of sodium carbonate (soda ash). CAUTION: Moderately toxic; irritant to mucous membranes. Wear rubber gloves when handling.

SALICYLIC ACID (Orthohydroxy Benzoic Acid): White powder, soluble in alcohol, oil of turpentine, and ether. Made by treating a hot solution of sodium phenolate with carbon dioxide.

SAND (Silicon Dioxide): White crystals or powder, soluble only in hydrofluoric acid or molten alkali. Found widely in its natural state. Can also be made from a soluble silicate (waterglass) by acidifying, washing, and igniting.

SASSAFRAS LEAVES: The leaves of the plant *Sassafras albidum.*

SESAME OIL (Benne Oil, Teel Oil): Bland yellow liquid, soluble in chloroform, ether, and carbon disulfide. Extracted from the plant *Sesamum indicum,* found in China, Japan, and South America.

SHELLAC (Garmet Lac, Gum Lac, Stick Lac): A natural resin secreted by the laccifer and deposited on trees in India. Soluble in alcohol.

SILICA GEL: Hard white lumps, crystals, or powder. Regenerative adsorbent, having a vast internal porosity in relation to its outside surface. Made by the reaction of sulfuric acid and sodium silicate.

SILICONE WATER EMULSION: A milky, slippery liquid that can be further diluted with water to any desired concentration. Made by the mixture of silicone oil, emulsifier, and water.

SILVER NITRATE: Transparent crystals, soluble in cold water or hot alcohol. Made by dissolving silver in dilute nitric acid and evaporating. CAUTION: Highly toxic. Strong irritant; handle with care.

SODA ASH (Sodium Carbonate): Grayish white powder, soluble in water. Mined in areas such as Great Salt Lake, or can be made by the Solvay ammonia soda process.

SODIUM ALGINATE: Colorless to light yellow solid. May be in granular or powdered form. Forms a thick collodial solution with water. Made by extraction from brown seaweed (kelp).

SODIUM ALUMINATE: White powder, soluble in water. Made

by heating bauxite with sodium carbonate and extracting the sodium aluminate with water.

SODIUM BICARBONATE (Baking Soda): White powder, soluble in water. Made by treating a saturated solution of soda ash with carbon dioxide.

SODIUM BISULFATE (Niter Cake): Colorless crystals soluble in water. A by-product in the manufacture of hydrochloric and nitric acids. CAUTION: Toxic in solution. Irritant to eyes and skin.

SODIUM CARBONATE: See Soda Ash.

SODIUM CHLORATE: Colorless, odorless crystals, soluble in water and alcohol. Made by heating and electrolyzing a concentrated solution of sodium chloride.

SODIUM CHLORIDE (Salt): White crystals soluble in water and glycerol. Made by the evaporation of salt brine.

SODIUM CITRATE: White crystals or powder, soluble in water. Made by treating a sodium sulfate solution with calcium citrate. CAUTION: Combustible.

SODIUM DIDELARO: White powder. Decomposes in water.

SODIUM DODECYLBENZENE SULFATE: White to light yellow flakes or powder. Biodegradable.

SODIUM HYDROXIDE: See Caustic Soda.

SODIUM HYPOCHLORITE: Pale greenish solution soluble in cold water. Made by the addition of chlorine to a cold dilute solution of sodium hydroxide. CAUTION: Toxic by ingestion and inhalation. Irritant to skin and eyes.

SODIUM LAURYL SULFATE: White or light yellow crystals, soluble in water. Acts as a wetting agent.

SODIUM METAPHOSPHATE: White powder, soluble in water.

SODIUM METASILICATE: A crystalline silicate. White granules, soluble in water.

SODIUM NITRATE (Saltpeter): Colorless, transparent crystals, soluble in water and glycerol. Made from nitric acid and sodium carbonate. CAUTION: Moderately toxic and flammable. Explodes when subjected to physical shock or temperatures of 1000° F.

SODIUM PENTACHLOROPHENATE: White to tan powder, soluble in water and acetone. CAUTION: Toxic by ingestion and inhalation. Irritant to eyes and skin.

SODIUM PERBORATE: White, odorless powder or crystals. Decomposes in water to release oxygen. Made by electrolysis of a solution of borax and soda ash.

SODIUM PHOSPHATE: White powder, soluble in water and alcohol. Made by precipitating calcium carbonate from a solution of dicalcium phosphate with soda ash.

SODIUM PYROPHOSPHATE: Transparent crystals or white powder. Soluble in water. Made by fusing sodium phosphate.

SODIUM SESQUICARBONATE (Sesqui): White needle-shaped crystals, soluble in water. Made by crystallation from a solution of sodium carbonate and sodium bicarbonate.

SODIUM SILICATE (Waterglass): Clear viscous liquid, soluble in water. Made by the fusion of sand and soda ash. CAUTION: May be irritating and caustic to skin and mucous membranes.

SODIUM SULFATE (Salt Cake): White crystals or powder, soluble in water. A by-product of hydrochloric acid production from salt and sulfuric acid.

SODIUM SULFITE: White crystals or powder, soluble in water.

Made by reacting sulfur dioxide with soda ash and water.

SODIUM THIOSULFATE (Hypo): White crystals or powder, soluble in water and oil of turpentine. Made by heating a solution of sodium sulfite with powdered sulfur.

SODIUM TRIPOLYPHOSPHATE (Tripoly): White powder, soluble in water. Made by calcination of sodium orthophosphate mixture from sodium carbonate and phosphoric acid.

SOFT SOAP: A liquid soap made with potassium hydroxide and a vegetable oil (except coconut and palm kernel oil).

SOYBEAN OIL (Chinese Bean Oil, Soy Oil): Pale yellow drying oil, soluble in alcohol, ether, chloroform, or carbon disulfide. Made by expression and solvent extraction of crushed soybeans.

SPERMACETI: White, semitransparent, waxlike solid, soluble in ether, chloroform, carbon disulfide, and hot alcohol. Extracted from the head of the sperm whale.

STANNIC OXIDE: White powder, soluble in sulfuric acid and hydrochloric acid. Found in nature in the mineral cassiterite.

STEARAMIDE: Colorless flakes, soluble in alcohol.

STEARIC ACID: Waxlike solid, soluble in alcohol, ether, chloroform, or carbon disulfide. Made by hydrogenation of oleic acid.

STODDARD SOLVENT: Water-white liquid solvent. CAUTION: Mildly flammable.

SUGAR: A carbohydrate product of photosynthesis comprising one, two, or more saccharose groups.

SULFITE LIQUOR: A waste liquor from the sulfite paper-making process. Synthetic vanilla (vanalin) is made from this material.

SULFITE LYE: See Sodium Hydroxide.

SULFONATED CASTOR OIL: A vegetable oil that has been treated with sulfuric acid and neutralized with a small amount of caustic soda. The oil is then emulsifiable with water.

SULFUR, WETTABLE: Pure sulfur exists in two stable crystalline forms. Alpha-sulfur: rhombic, octahedral, yellow crystals stable at room temperature. Beta-sulfur: monoclinic, prismatic, pale yellow crystals. Both forms are insoluble in water; slightly soluble in alcohol and ether; soluble in carbon disulfide, carbon tetrachloride, and benzene.

SULFURIC ACID (Hydrogen Sulfate, Oil of Vitroil, Battery Acid): Strongly corrosive, dense, oily liquid, colorless to dark brown, depending on purity. Miscible with water in all proportions. Dissolves most metals; concentrated acid oxidizes, dehydrates, or sulfonates most organic compounds. CAUTION: Use caution in mixing with water. Always add the acid to water, *never* the reverse. Highly toxic and strong irritant to tissue. Wear gloves and use in well-ventilated area.

SULFONATED CASTOR OIL: A vegetable oil that has been treated with sulfuric acid and neutralized with a small amount of caustic soda. The oil is then emulsifiable with water.

SUPERPHOSPHATE (Acid Phosphate): Water-soluble powder, made by the action of sulfuric acid on insoluble rock.

TALC (Talcum, Mineral Graphite, Steatite): A mined mineral (magnesium silicate), white-gray pearly color with a greasy feel.

TANNIC ACID (Gallotainic Acid, Tannin): Light yellow crystals, soluble in water, alcohol, and benzene. Made by extraction of nutgalls and tree bark, with water and alcohol.

TAPIOCA FLOUR (Amylum): White amorphous, tasteless powder; irregular lumps or fine powder. Insoluble in cold water, alcohol, and ether; forms a gel with hot water. Derived from cassava (tapioca).

TETRASODIUM PYROPHOSPHATE (TSP): Colorless crystals or white powder, soluble in water. Made by fusing sodium phosphate.

TARTARIC ACID: White crystalline powder, soluble in water and alcohol. Made from maleic anhydride and hydrogen peroxide.

TIN, POWDERED (STANNUM): White ductile solid. Metallic element of atomic number 50; group IVA of the periodic system.

TINCTURE OF ARNICA: Medication derived from a plant, usually in tinctures of various strengths. CAUTION: Toxic by ingestion.

TINCTURE OF BENZOIN: Clear to pale yellow liquid, having a slight camphor odor. The crystals from which the tincture is made are derived from the condensation of benzaldehyde in a cyanide solution.

TINCTURE OF IODINE: Water and alcohol mixture of potassium iodine used medically as an antiseptic. CAUTION: Concentration of iodine may increase as alcohol evaporates. Keep tightly capped, avoiding older containers that have been previously opened; iodine can cause severe burns in its concentrated forms.

TINCTURE OF RHUBARB: Dried root and stalks of rhubarb are treated with alcohol to form a tincture (about 10% solution).

TITANIUM DIOXIDE (Titanium White, Titania): White powder, miscible with water, alcohol, or oil. Made by treating ilmenite with sulfuric acid.

TITANIUM TRICHLORIDE: Dark violet crystals, soluble in alcohol. Flammable in the presence of oxidizing materials. CAUTION: Skin irritant.

TOLUENE (Toluol, Methylbenzene): White liquid, soluble in alcohol, benzene, and ether. Madc by fractional distillation of coal tar oil. CAUTION: Flammable.

TRAGACANTH (Tragacanth Gum): White flakes or yellow powder. Soluble in alkaline solution.

TRICHLOROETHYLENE (Tri, Trichlor): Colorless, heavy liquid; slightly soluble in water, miscible with organic solvents. Made from tetrachloroethane by treatment with alkali in the presence of water. CAUTION: Vapors are toxic. Use with adequate ventilation.

TRIETHANOLAMINE (TEA, TRI): Colorless viscous hygroscopic liquid, miscible with water and alcohol. Made by the reaction of ethylene oxide and ammonia. CAUTION: May be somewhat irritating to skin and mucous membranes.

TRISODIUM PHOSPHATE (Sodium Phosphate Dibasic): Colorless crystals or white powder, soluble in water and alcohol. Made by precipitating calcium carbonate from a solution of dicalcium phosphate with soda ash. CAUTION: Skin irritant, use rubber gloves. Moderately toxic by ingestion.

TURPENTINE: Colorless, clear, oily liquid. Made by steam distillation of turpentine gum. CAUTION: Toxic if taken internally. Flammable. Handle with care.

ULTRAMARINE BLUE: Blue lumps, soluble in oil. Made by heating a mixture of sulfur, clay, alkali, and reducer at high temperatures.

UREA (Carbamide): White crystals or powder, soluble in water, alcohol, and benzene. Derivation: liquid ammonia and liquid carbon dioxide react under pressure and elevated temperatures to form ammonium carbonate, which decomposes at lower temperatures to form urea.

UREA PHOSPHATE: See Urea.

VENICE TURPENTINE (Larch Gum): Yellow to greenish resin, soluble in most organic solvents. Distilled from *Larix europaca.*

VERMICULITE: Crystalline-type structure with high porosity.

Insoluble, except in hot acids. Used as a filler in concrete and for thermal insulation.

VINEGAR (Dilute Acetic Acid): Brown liquid, dilutable with water. Made by fermentation of fruits and grains. May be distilled to remove brown color, after which it is known as white vinegar.

WHITE BEESWAX: Wax from the honeycomb of frames in the beehive. White color is obtained by bleaching the natural yellow wax. Soluble in chloroform, ether, and oils. Melting point 62°-65° C.

WITCH HAZEL: A clear white, astringent liquid, soluble in water and alcohol.

WOOD TAR (Pine Tar): Viscous, sticky, brown-to-black syrup, soluble in alcohol and acetone. Made by the destructive distillation of pine wood.

XYLOL (Xylene): Clear liquid, soluble in alcohol and ether. Made by fractional distillation of petroleum, coal tar, or coal gas. CAUTION: Flammable.

YELLOW BEESWAX: See White Beeswax. Note: Yellow and white beeswax have the same properties except color. Therefore, where color is not important (as in floor wax, for example), the yellow wax is more economical.

YELLOW DEXTRIN: Yellow or white powder, soluble in water. Made by heating dry starch or by treating starch with dilute acid.

ZINC: Shining white metal of atomic number 30. Soluble in acids and alkalies. Mined in British Columbia, Utah, Colorado, Idaho, Peru, and Australia.

ZINC ACETATE: White crystalline plates, soluble in water and alcohol. Made by the action of acetic acid on zinc oxide.

ZINC BROMIDE: White crystalline powder, soluble in water,

alcohol, and ether. Made by the interaction of solutions of barium bromide and zinc sulphate, and then crystallized.

ZINC CHLORIDE: White crystals or crystalline powder, soluble in water, alcohol, and glycerin. Made by the action of hydrochloric acid on zinc. CAUTION: Toxic.

ZINC OXIDE (Chinese White, Zinc White): Coarse white to gray powder, soluble in acids and alkalies. Made by oxidation of vaporized pure zinc. CAUTION: Poisonous if taken internally.

ZINC STEARATE: Soluble in common solvents and hot acids. Made by the action of sulfuric acid on zinc.

ZINC SULFATE (White Vitriol): Colorless crystalline powder, soluble in water or glycerol. Made by the action of sulfuric acid on zinc oxide.

# APPENDIX C

# Sources of Chemical Supplies

The majority of ingredients in the formulas of this book are easily obtained from local sources in most areas of the United States. Borax, mineral oil, kerosene, beeswax, lanolin, paraffin wax, salt, soda ash, baking soda, and glycerin are repeatedly called for in formulating the products; most of these ingredients are in the local drug, grocery, paint, or fuel supply stores. Most of the others can be obtained from specialized stores, manufacturers, or chemical supply houses.

While it is suggested that chemicals be purchased from local sources wherever possible, some materials are not as common as others and are more difficult to locate, especially in nonmetropolitan areas. If this problem occurs or if you prefer to order by mail at prices usually lower than those in retail stores, there are two alternatives: you can (1) buy your materials by mail from a chemical repackager, who buys chemicals in bulk lots and repackages them in small quantities, or (2) contact the manufacturer, who can advise you of local sources of supply for the products it manufactures.

Many chemical manufacturers have offices in principal cities. Frequently, they are listed under the product you wish to locate. Also, principal manufactures are listed, under the product you seek, in the *O.P.D. Chemical Buyers' Directory* (Schuell Publishing Company). Most libraries have a copy.

Below is a list of ingredients and sources of supplies used in *The Formula Book 3*. The number which follows the source represents the number of times that compound or chemical is called for in formulating products. They are listed as a convenience to the consumer, who may use the information when purchasing the ingredients. For example, glycerin is used in 18 formulas; it would therefore be foolish to purchase a small quantity of the material when it is called for so many times.

ACACIA: Drugstore, drug distributor, industrial chemical supplier. (1)

ALMOND OIL: Retail and wholesale drugstores. (1)

ALUMINUM POTASSIUM SULFATE: Retail and wholesale drugstores, industrial chemical suppliers. (3)

ALUMINUM POWDER: Retail and wholesale paint and hardware stores, industrial chemical suppliers. (1)

ALUMINUM SULFATE: Retail and wholesale drugstores, industrial chemical suppliers, ceramic and hobby shops, repackagers of chemicals. (1)

AMMONIA, HOUSEHOLD: Retail supermarkets, wholesale groceries, feed and grain suppliers. (1)

AMMONIUM CARBONATE: Retail and wholesale drugstores, industrial chemical suppliers. (3)

AMMONIUM CHLORIDE: Retail and wholesale drugstores, industrial chemical suppliers. (2)

AMMONIUM CITRATE: Retail and wholesale drugstores, industrial chemical suppliers. (1)

AMMONIUM STEARATE: Retail and wholesale drugstores, industrial chemical suppliers, repackagers of chemicals. (2)

AMMONIUM SULFATE: Retail and wholesale drugstores, industrial chemical suppliers, repackagers of chemicals. (2)

AMYL ACETATE: Wholesale paint and hardware stores, industrial chemical suppliers, repackagers of chemicals. (1)

ANTIMONY CHLORIDE: Wholesale paint and hardware stores, industrial chemical suppliers, repackagers of chemicals. (1)

ASPHALT: Retail and wholesale paint and hardware stores, building material and supply dealers. (1)

BENTONITE: Retail paint and hardware stores, industrial chemical suppliers, building-supply dealers. (1)

BENZENE: Retail and wholesale paint and hardware stores, oil distributors, industrial chemical suppliers, repackagers of chemicals. (1)

BORAX: Retail and wholesale paint and hardware stores, retail supermarkets, wholesale groceries, industrial chemical suppliers, repackagers of chemicals. (3)

BORIC ACID: Retail and wholesale drugstores, industrial chemical suppliers, repackagers of chemicals. (7)

BUTYL ALCOHOL: Retail and wholesale drugstores, industrial chemical suppliers, repackagers of chemicals. (1)

CALCIUM CARBONATE: Retail and wholesale drugstores, industrial chemical suppliers, feed and grain suppliers, building materials and supply dealers, repackagers of chemicals. (4)

CARNAUBA WAX: Oil distributors, industrial chemical suppliers, repackagers of chemicals. (1)

CARBOLIC ACID: Retail and wholesale drugstores, industrial chemical suppliers, repackagers of chemicals. (2)

CARBON BLACK: Retail paint and hardware stores, industrial chemical suppliers, ceramic and hobby shops, repackagers of chemicals. (3)

CASTOR OIL: Retail and wholesale drugstores, industrial chemical suppliers, repackagers of chemicals. (3)

CAUSTIC POTASH: Retail paint and hardware stores, industrial chemical suppliers, ceramic and hobby shops, repackagers of chemicals. (2)

CAUSTIC SODA: Retail and wholesale drugstores and paint and hardware stores, industrial chemical suppliers, repackagers of chemicals. (4)

CERESIN WAX: Industrial chemical suppliers, ceramic and hobby shops, repackagers of chemicals. (2)

CHLOROPHYLL: Retail and wholesale drugstores, industrial chemical suppliers, repackagers of chemicals. (1)

CITRIC ACID: Retail and wholesale drugstores, industrial chemical suppliers, repackagers of chemicals. (3)

COCONUT OIL: Retail and wholesale drugstores, industrial chemical suppliers, repackagers of chemicals. (2)

COPPER NAPHTHENATE: Retail and wholesale drugstores, industrial chemical suppliers, repackagers of chemicals. (2)

CREOSOTE: Industrial chemical suppliers, building materials and supply dealers, repackagers of chemicals. (2)

DIATOMACEOUS EARTH: Industrial chemical suppliers, building materials and supply dealers, repackagers of chemicals, ceramic and hobby shops. (1)

FERRIC CHLORIDE: Retail and wholesale drugstores, industrial chemical suppliers, repackagers of chemicals. (1)

FISH OIL: Retail and wholesale paint and hardware stores, feed and grain suppliers. (1)

FLOWERS OF SULFUR: Retail and wholesale drugstores, industrial chemical suppliers, repackagers of chemicals. (1)

FORMALDEHYDE: Retail and wholesale drugstores, industrial chemical suppliers, repackagers of chemicals. (1)

FULLER'S EARTH: Industrial chemical suppliers, building materials and supply dealers, ceramic and hobby shops, repackagers of chemicals. (2)

GLYCERIN: Retail and wholesale drugstores, industrial chemical suppliers. (18)

HYDROGEN PEROXIDE: Retail and wholesale drugstores, industrial chemical suppliers, repackagers of chemicals. (3)

LANOLIN: Retail and wholesale drugstores, industrial chemical suppliers, repackagers of chemicals. (8)

LEMON OIL: Retail and wholesale drugstores, industrial chemical suppliers, repackagers of chemicals. (2)

LINSEED OIL: Retail and wholesale paint and hardware stores, industrial chemical suppliers, repackagers of chemicals. (2)

MAGNESIUM STEARATE: Retail and wholesale drugstores, industrial chemical suppliers, oil distributors, repackagers of chemicals. (1)

MAGNESIUM SULFATE: Retail and wholesale drugstores, industrial chemical suppliers, repackagers of chemicals. (1)

MENTHOL: Retail and wholesale drugstores, industrial chemical suppliers, repackagers of chemicals. (1)

METHYL SALICYLATE: Retail and wholesale drugstores, industrial chemical suppliers, repackagers of chemicals. (1)

MINERAL OIL, WHITE: Retail and wholesale drugstores, service stations, oil distributors, industrial chemical suppliers, repackagers of chemicals. (4)

MINERAL SPIRITS: Service stations, oil distributors, industrial chemical suppliers, repackagers of chemicals. (1)

NAPHTHA: Oil distributors, industrial chemical suppliers, repackagers of chemicals. (1)

NEATSFOOT OIL: Feed and grain suppliers, oil distributors, repackagers of chemicals. (3)

OLEIC ACID: Retail and wholesale drugstores, industrial chemical suppliers, repackagers of chemicals. (4)

OXALIC ACID: Retail and wholesale drugstores, industrial chemical suppliers, repackagers of chemicals. (1)

PARAFFIN OIL: Service stations, oil distributors, industrial chemical suppliers, repackagers of chemicals. (2)

PARAFFIN WAX: Retail supermarkets, wholesale groceries, oil distributors, industrial chemical suppliers, repackagers of chemicals. (5)

PEANUT OIL: Retail supermarkets, wholesale groceries, oil distributors, repackagers of chemicals. (1)

PEPPERMINT OIL: Retail and wholesale drugstores, repackagers of chemicals. (3)

PETROLATUM: Retail and wholesale drugstores, oil distributors, industrial chemical suppliers, repackagers of chemicals. (5)

PINE OIL: Retail and wholesale drugstores, industrial chemical suppliers, repackagers of chemicals. (3)

POTASSIUM OLEATE: Retail and wholesale drugstores, industrial chemical suppliers, repackagers of chemicals. (1)

PUMICE POWDER: Retail and wholesale drugstores, industrial chemical suppliers, repackagers of chemicals. (2)

QUASSIA: Retail and wholesale drugstores, repackagers of chemicals. (1)

SESAME OIL: Retail and wholesale drugstores, industrial chemical suppliers, repackagers of chemicals. (1)

SHELLAC: Retail and wholesale paint and hardware stores, industrial chemical suppliers, repackagers of chemicals. (1)

SILICONE WATER EMULSION: Repackagers of chemicals, foundry suppliers. (3)

SODA ASH: Retail and wholesale paint and hardware stores, industrial chemical suppliers, repackagers of chemicals. (6)

SODIUM ALUMINATE: Retail and wholesale drugstores, industrial chemical suppliers, repackagers of chemicals. (1)

SODIUM BICARBONATE: Retail and wholesale paint and hardware stores, retail supermarkets, repackagers of chemicals. (5)

SODIUM CHLORIDE: Retail supermarkets, wholesale groceries, feed and grain suppliers. (2)

SODIUM METAPHOSPHATE: Retail and wholesale drugstores, industrial chemical suppliers, repackagers of chemicals. (1)

SODIUM METASILICATE: Retail and wholesale drugstores, industrial chemical suppliers, repackagers of chemicals. (1)

SODIUM PERBORATE: Retail and wholesale drugstores, industrial chemical suppliers, repackagers of chemicals. (3)

SODIUM PYROPHOSPHATE: Oil distributors, industrial chemical suppliers, repackagers of chemicals. (1)

SODIUM SESQUICARBONATE: Retail and wholesale drugstores, industrial chemical suppliers, repackagers of chemicals. (1)

SODIUM SILICATE: Retail and wholesale drugstores, industrial chemical suppliers, repackagers of chemicals. (2)

STANNIC OXIDE: Retail and wholesale drugstores, industrial chemical suppliers, oil distributors, repackagers of chemicals. (2)

STEARIC ACID: Retail and wholesale drugstores, oil distributors, industrial chemical suppliers, repackagers of chemicals. (8)

STODDARD SOLVENT: Retail drugstores, oil distributors, repackagers of chemicals, dry cleaners' suppliers. (3)

SULFONATED CASTOR OIL: Retail and wholesale drugstores, industrial chemical suppliers, oil distributors, repackagers of chemicals. (1)

TALC: Retail and wholesale drugstores, oil distributors, repackagers of chemicals. (1)

TANNIC ACID: Retail and wholesale drugstores, oil distributors, industrial chemical suppliers, repackagers of chemicals. (4)

TARTARIC ACID: Retail and wholesale drugstores, industrial chemical suppliers, repackagers of chemicals. (1)

TINCTURE OF BENZOIN: Retail and wholesale drugstores, industrial chemical suppliers, repackagers of chemicals. (2)

TITANIUM DIOXIDE: Retail and wholesale drugstores, oil distributors, repackagers of chemicals. (1)

TRAGACANTH: Retail and wholesale drugstores, industrial chemical suppliers, repackagers of chemicals. (2)

TRIETHANOLAMINE: Retail and wholesale drugstores, industrial chemical suppliers, repackagers of chemicals. (10)

TRISODIUM PHOSPHATE: Retail and wholesale paint and hardware stores, industrial chemical suppliers, repackagers of chemicals. (8)

TURPENTINE: Retail and wholesale paint and hardware stores, industrial chemical suppliers. (2)

WHITE BEESWAX: Retail and wholesale drugstores, industrial chemical suppliers, repackagers of chemicals. (6)

ZINC OXIDE: Retail and wholesale drugstores, industrial chemical suppliers, repackagers of chemicals. (3)

The following companies handle the ingredients listed in the formulas of this book and are presented as a convenience to our readers. No endorsement is given or implied for any company.

*CHEMICALS: Student Science Service, 622 W. Colorado St., Glendale, Calif. 91204, (213)-247-6910.*

*PERFUMES AND ESSENTIAL OILS: Sherman Toy Corporation, P.O. Box 455 WOB, West Orange, N.J. 07052.*

APPENDIX D

# Utensils and Equipment

The formulas in *The Formula Book 3* are designed for small volumes of the finished product, requiring minimum equipment to formulate. The following utensils are required.

1. Several glass measuring cups (see Figure 2).
2. A set of mixing bowls made of glass, ceramic or plastic (see Figure 10).
3. A wood fork with spacing of about ⅛ inch, between tines (see Figure 3).
4. An egg beater (see Figure 4). An electric mixer, with beaters and bowl, is helpful but not essential. If it has variable speeds, it can be used for both wet and dry mixing, saving a great deal of time and assuring a "good blend."
5. A stem-type thermometer is convenient, but again, not absolutely essential (see Figure 5). If one is not available, remember that water gives off a mild vapor at 140° F., a moderate vapor at 160° F., a heavy vapor at 180° F., and heavy steam at the boiling point.
6. A supply of wood tongue depressors (see Figure 8). These are smooth, cheap, and readily available from any druggist. They make excellent mixing sticks, and are inexpensive enough to be disposable, eliminating a lot of "cleaning up."
7. Paper cups are ideal for small-batch formulating. They are inexpensive, disposable, and can be easily numbered or marked with a felt marker.
8. Double boilers are required in many instances. These should be Pyrex (see Figure 6).
9. A rubber syringe for measuring drops (see Figure 7).
10. A set of standard measuring spoons (see Figure 11a).
11. A plastic cone and filter paper, such as used in coffee

making (see Figure 9). While filtering a liquid compound after it is finished is not usually essential, it is always desirable, in that a clearer, better-looking product results.

12. Containers for the finished product are a matter of personal preference. In most homes, jars and bottles are available. If they are to be purchased, many supermarkets carry them, and drugstores have them for their own use. Larger quantities can be had from bottle distributors, listed in the Yellow Pages.

*All* chemicals that are stored in containers should be labeled, regardless of whether they are a raw material or a finished compound. This is basic, and *must* be followed in the interest of safety. Keep all chemicals out of the reach of children, and note the contents on the label. In this way if a child, or even an unsuspecting adult, should accidentally consume the contents, the doctor would know what treatment to initiate. While these formulas have been chosen with an eye to safety, many materials normally regarded as safe can be dangerous if taken internally, or to excess. Here's an example of a safe label.

**This Bottle Contains** ______________________
**Its Ingredient(s) are:**

**Keep out of reach of children.**

**Made by** ______________________

**Date** ______________________

***KEEP BOTTLE SEALED***

# Illustrations

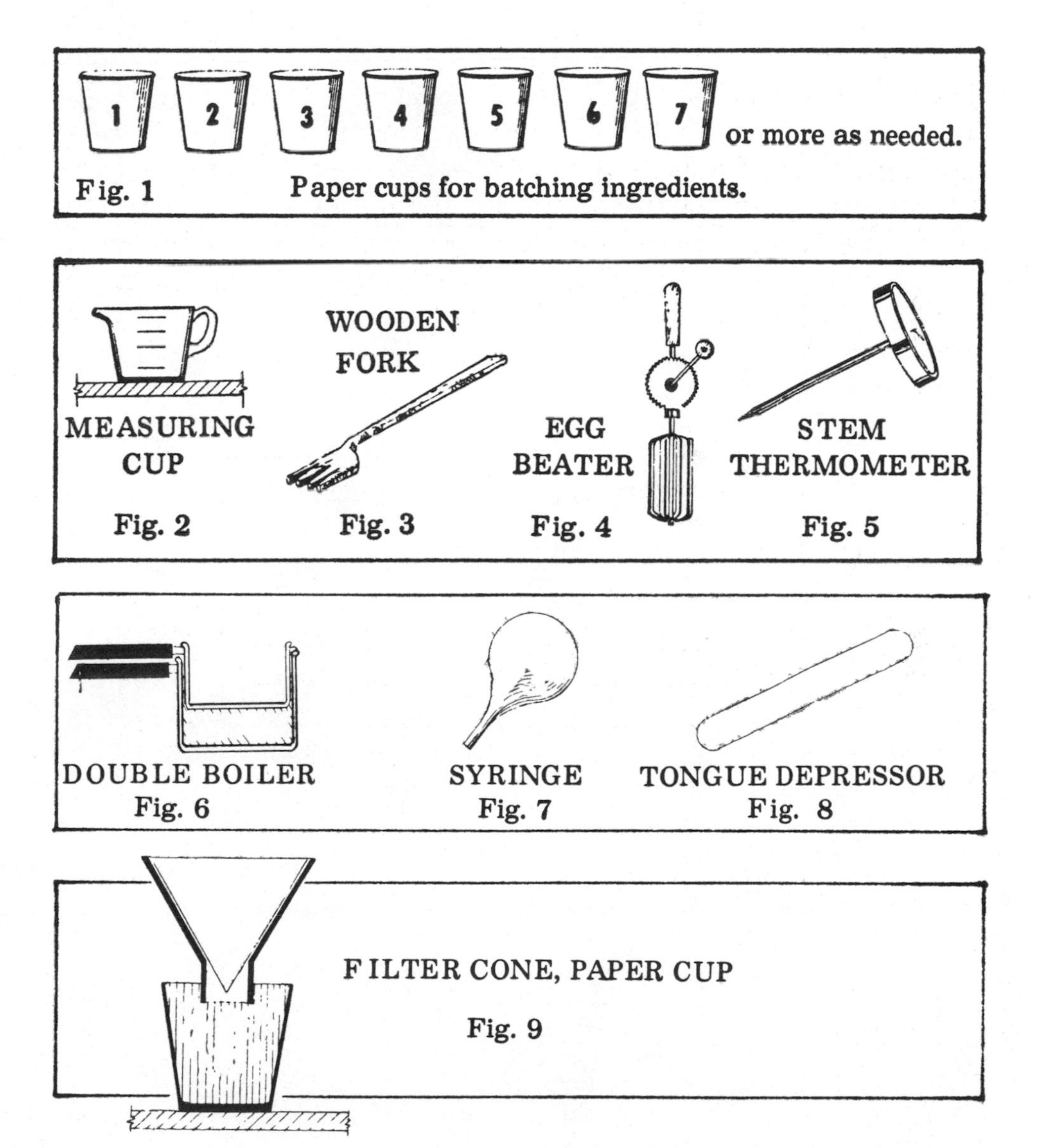


Fig. 1 Paper cups for batching ingredients.

Fig. 2 Fig. 3 Fig. 4 Fig. 5

Fig. 6 Fig. 7 Fig. 8

Fig. 9

MIXING BOWLS

Fig. 10

SPECK MEASURE

Fig. 12a

FILLING

Fig. 11

MEASURING SOLIDS

Fig. 12 Fig. 13

MEASURING SPOONS

11a

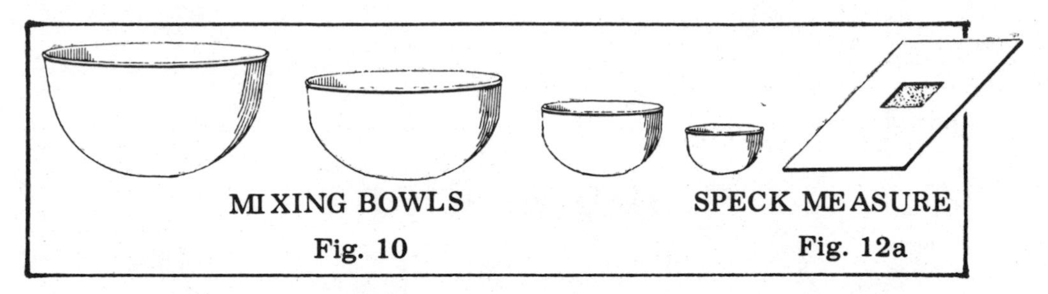

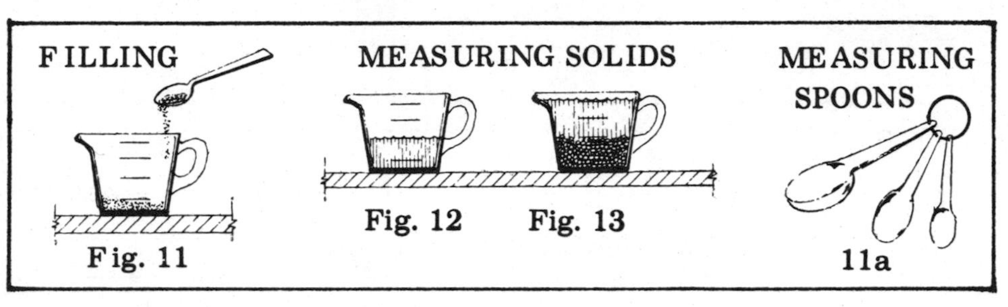

This Bottle Contains ______
Its Ingredient(s) are:

Keep out of reach of children.

Made by ______

Date ______

*KEEP BOTTLE SEALED*

This Bottle Contains ______
Its Ingredient(s) are:

Keep out of reach of children.

Made by ______

Date ______

*KEEP BOTTLE SEALED*

This Bottle Contains ______
Its Ingredient(s) are:

Keep out of reach of children.

Made by ______

Date ______

*KEEP BOTTLE SEALED*

This Bottle Contains ______
Its Ingredient(s) are:

Keep out of reach of children.

Made by ______

Date ______

*KEEP BOTTLE SEALED*

This Bottle Contains ______
Its Ingredient(s) are:

Keep out of reach of children.

Made by ______

Date ______

*KEEP BOTTLE SEALED*

This Bottle Contains ______
Its Ingredient(s) are:

Keep out of reach of children.

Made by ______

Date ______

*KEEP BOTTLE SEALED*

This Bottle Contains ______
Its Ingredient(s) are:

Keep out of reach of children.

Made by ______

Date ______

*KEEP BOTTLE SEALED*

This Bottle Contains ______
Its Ingredient(s) are:

Keep out of reach of children.

Made by ______

Date ______

*KEEP BOTTLE SEALED*

This Bottle Contains ______
Its Ingredient(s) are:

Keep out of reach of children.

Made by ______

Date ______

*KEEP BOTTLE SEALED*

This Bottle Contains ______
Its Ingredient(s) are:

Keep out of reach of children.

Made by ______

Date ______

*KEEP BOTTLE SEALED*

This Bottle Contains ______
Its Ingredient(s) are:

Keep out of reach of children.

Made by ______

Date ______

*KEEP BOTTLE SEALED*

This Bottle Contains ______
Its Ingredient(s) are:

Keep out of reach of children.

Made by ______

Date ______

*KEEP BOTTLE SEALED*

**This Bottle Contains** ____________________
**Its Ingredient(s) are:**

**Keep out of reach of children.**

**Made by** ____________________

**Date** ____________________

***KEEP BOTTLE SEALED***

**This Bottle Contains** ____________________
**Its Ingredient(s) are:**

**Keep out of reach of children.**

**Made by** ____________________

**Date** ____________________

***KEEP BOTTLE SEALED***

**This Bottle Contains** ____________________
**Its Ingredient(s) are:**

**Keep out of reach of children.**

**Made by** ____________________

**Date** ____________________

***KEEP BOTTLE SEALED***

**This Bottle Contains** ____________________
**Its Ingredient(s) are:**

**Keep out of reach of children.**

**Made by** ____________________

**Date** ____________________

***KEEP BOTTLE SEALED***

**This Bottle Contains** ____________________
**Its Ingredient(s) are:**

**Keep out of reach of children.**

**Made by** ____________________

**Date** ____________________

***KEEP BOTTLE SEALED***

**This Bottle Contains** ____________________
**Its Ingredient(s) are:**

**Keep out of reach of children.**

**Made by** ____________________

**Date** ____________________

***KEEP BOTTLE SEALED***

**This Bottle Contains** ____________________
**Its Ingredient(s) are:**

**Keep out of reach of children.**

**Made by** ____________________

**Date** ____________________

***KEEP BOTTLE SEALED***

**This Bottle Contains** ____________________
**Its Ingredient(s) are:**

**Keep out of reach of children.**

**Made by** ____________________

**Date** ____________________

***KEEP BOTTLE SEALED***

**This Bottle Contains** ____________________
**Its Ingredient(s) are:**

**Keep out of reach of children.**

**Made by** ____________________

**Date** ____________________

***KEEP BOTTLE SEALED***

**This Bottle Contains** ____________________
**Its Ingredient(s) are:**

**Keep out of reach of children.**

**Made by** ____________________

**Date** ____________________

***KEEP BOTTLE SEALED***

**This Bottle Contains** ____________________
**Its Ingredient(s) are:**

**Keep out of reach of children.**

**Made by** ____________________

**Date** ____________________

***KEEP BOTTLE SEALED***

**This Bottle Contains** ____________________
**Its Ingredient(s) are:**

**Keep out of reach of children.**

**Made by** ____________________

**Date** ____________________

***KEEP BOTTLE SEALED***

NOTES

Name of Formula: ____________________

Date Made: ____________

Ingredients and amounts: ____________________

____________________

____________________

____________________

Label: Ingredients and caution warnings

Observations: ____________________

____________________

____________________

____________________

____________________

----------

Name of Formula: ____________________

Date Made: ____________

Ingredients and amounts: ____________________

____________________

____________________

____________________

Label: Ingredients and caution warnings

Observations: ____________________

____________________

____________________

____________________

____________________

NOTES

Name of Formula: ____________________

Date Made: ____________

Ingredients and amounts: ____________________

____________________

____________________

____________________

Label: Ingredients and caution warnings

Observations: ____________________

____________________

____________________

____________________

____________________

----------

Name of Formula: ____________________

Date Made: ____________

Ingredients and amounts: ____________________

____________________

____________________

____________________

Label: Ingredients and caution warnings

Observations: ____________________

____________________

____________________

____________________

____________________

# Index